T0090812

CYCLES OF TIME

FROM INFINITY TO ETERNITY

Order this book online at www.trafford.com
or email orders@trafford.com

Most Trafford titles are also available at major online book retailers.

Printed in the United States of America.

ISBN: 978-1-4269-5336-1 (sc)
ISBN: 978-1-4269-5337-8 (hc)
ISBN: 978-1-4269-5349-1 (e)

Library of Congress Control Number: 2010919283

Trafford rev. 01/06/2011

www.trafford.com

North America & international
toll-free: 1 888 232 4444 (USA & Canada)
phone: 250 383 6864 • fax: 812 355 4082

To those who conquer it

Contents

Acknowledgments

My sincere thanks to my colleague and amateur astronomer, Thomas Flöck, for his careful review and for drawing my attention to several fascinating time related astronomical facts.

I am indebted to my philosophical soul mates and reviewers of my earlier books, Olivier Schreiber and Gene Poole, who again significantly contributed to the clarity of the exposition by pointing out mistakes and proposing corrections.

Special thanks are due to Attila Jeney for the quality figures and Rodrigo Adolfo for the cover art. I also appreciate the contributions of Nika Corales coordinator and Anna Mills production assistant at Trafford Publishing.

Finally, as always when writing in this genre, my wife Gabriella was the first sounding board. She deserves credit for asking questions the readers might also ask and enabling me to answer them in advance.

Spring equinox, 2011
Louis Komzsik

Prologue

Time. One of life's most enigmatic aspects. We are acutely aware of its constant presence, its passing and irreversible nature. We foolishly attempt to influence it. We rush it when waiting for some exciting event of our life, a birth, a wedding, or simply a vacation. Then we would like to reverse it and relive some joyous moments of the past.

No matter how hard we try, we can't change it. It is a steadily moving locomotive on the rails of our lives, its inertia beyond our means to influence. Or is it? Can we slow it down, stop it, or even reverse it?

We will try to understand these possibilities in this book that is a travel of a sort, not in time, but around time. The chapters mimic a year of human understanding of the phenomenon. The fact that we will reach December does not mean we fully understand time, far from it. It only means that we are through a cycle in our quest to understand time.

First we will explore the history of time, its origins in the celestial cycles observed by ancient cultures and expand our time horizon into cosmic cycles. We will examine and debunk the doomsday predictions about the year 2012.

The large scale measuring of time was gradually refined by humankind with medium scale instruments of calendars that brought time to the horizon of a human lifetime. The final, smaller scale measurements, via clocks of our everyday life, also bring an intriguing story as we will see.

Then we will investigate the more difficult physical aspects of time, first contributing to a more accurate measurement. We will recognize time's relativity and ponder the possibility of time travel. Intriguing conundrums lie on that road, including that of the twins' dilemma and the idea of stopping time. Influencing one's past, an attractive idea to all of us and leading to the grandfather paradox, may not be theoretically possible, but is worthy of an examination.

Finally we will attempt to visualize time's phases using a geometric infinity to present a temporal infinity, eternity. We will contemplate eternity, consider the faint hopes of extending our lives to it.

1

January: Celestial cycles

The high priest reached the flat platform on the top of the pyramid just as total darkness arrived. He could not see the town on the ground and the sky was also black apart from the many shiny objects, the subjects of his intense scrutiny for most of his adult life. The Great Star was bright and the familiar celestial landscape comforted him as he started his nightly observations.

He opened the ancient notes of his predecessors in the high priest role and glanced at them by the faint light of the Night Guard. Then he observed the sky for a long time. He had a feeling that the celestial alignments were now exactly the same as was recorded by someone many-many lifetimes before him. He looked at the records and validated his finding.

He pondered the significance of that for a while until he realized that the events in the sky will start a new cycle. Much like the daily cycles between the Day God and the Night Guard, or the longer duration changes in the path of the Day God that resulted in the seasons and weather changes. This cycle, however, appeared to be of a tremendous length of time, assuming the record's dates were to be trusted. Soon comes the day when the two children of the Day God, the Great Star and the Night Guard will visit their father again.

He carefully recorded his own observation and decided to finish for the night. After descending the many steps to the village ground below, he will start a new cycle of the calendar of celestial events for his successors in future generations. The new cycle will last until the spectacular phenomenon of both the Great Star and the Night Guard visiting the Day God will happen again.

This could have been the time of the beginning of the current and last Mayan calendar in 3114 BC.

Mayans are considered to be descendants of a wave of humans that crossed the Bering Strait around 20,000 years ago. Following a path along the Pacific, they crossed into the Yucatan peninsula and settled. Their mythology describes a person of great abilities, who led their migration from the other side of the ocean on a mysterious open path. Ice age land bridge assumptions of modern archeology seem to rhyme with this ancient belief.

Their Yucatan kingdom founder, Kukulkan was considered to be the descendant of the Sun, their Day God. His organization of the nation into four tribes and their four tribal centers is supported by archeological evidence. One of those centers, Chichen-Itza is substantially restored and is a testament to some of the Mayan records. Their kingdom lasted well into the middle ages and was ultimately destroyed in the middle of the 15th century by the Spanish colonists.

They were skilled astronomers, experts in observing the Moon, the Night Guard of the ancient priests. The Moon's cycle was a subject of high importance. Because of the lack of artificial light, Moon had a big effect on everyday life by extending the daylight hours. Even on lower latitudes, like those of the Mayan territory where the daylight hours were longer and their seasonal change was less notable, the reliance on Moon was very strong.

Moon has continuously changing and very distinct shapes between its phases. The elapsed time between two full Moon phases became the lunar cycle, the first and probably most influential celestial cycle. There was a less noticeable change between the full Moons, their size. Since Moon rotates around Earth on an elliptic orbit, the full Moon occurring when Moon is closest to Earth (called perigee) is larger than the one occurring when Moon is the farthest (called apogee).

The result of this was that the lunar cycles changed and the Mayans noticed that. They even had a different hieroglyph for them. Interestingly they counted their lunar months in units of six, numbered from 0 to 5. Having a zero at such early period was remarkable, considering the fact that the advanced Mediterranean society of the Greeks still struggled with its recognition as a valid mathematical constant two millennia later.

The six lunar months may have been related to another celestial cycle that the ancient Mayans also intensely observed, the eclipses of Moon. They recognized a distinction between the two types of eclipses.

A solar eclipse occurs when Moon passes in front of the Sun and when viewed from a certain region on the Earth, completely covers it. The other eclipse type, lunar eclipse occurs when Moon passes through the shadow of the Earth, which is visible from the night half of the Earth.

The probability of such celestial event is extremely low. Let us consider some numbers: The diameter of the Sun is approximately 864,000 miles and its average distance from us is about 93 million miles. The same numbers for the Moon are: diameter 2160 miles and its average distance is 238,000 miles, a stunning demonstration of the spectacular achievement of the Apollo astronauts landing on the Moon.

It is well known in everyday circumstances that an object at a distance appears smaller and smaller as the distance increases. Furthermore, a smaller object that is closer than a larger one may appear the same size, a fact exploited by artists in medieval times. The apparent sizes of two objects are based on the ratio of their actual sizes and their distances. This is the concept at work, when we cover the Moon with our thumb in front of our eyes.

These numbers for our celestial objects are the following. The ratio of their sizes is $864000/2160 \approx 400$. The ratio of their distances is $93000000/238000 \approx 390$. A most remarkable coincidence! The ratio by which the Sun is larger than the Moon is the same as the ratio by which the Moon is closer to Earth than the Sun, resulting in Moon's ability to cover the Sun during eclipses. The earthly visual angle of the Sun and the Moon is the same, about half a degree.

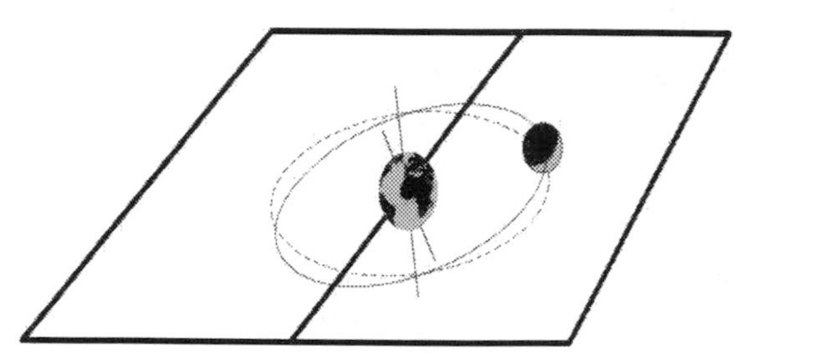

FIGURE 1.1 Moon's orbit

Because of the angle between the plane of the Moon orbiting Earth and the plane of Earth's orbit around the Sun, shown on Figure 1.1, the eclipses occur at irregular, albeit predictable times. On the figure the solid circle depicts Moon's orbit around the Earth, while the dotted line is the projection of Moon's orbit to Earth's orbital plane represented by the parallelogram on the figure. Most of the time the Moon passes above or below the plane of Earth's orbit, and eclipses could only occur when Moon's orbit crosses the plane of Earth's orbit.

Hence eclipses could only occur twice a year, approximately six months apart, that fact could also be the reason for the Mayans counting from zero to five. It appears that the Mayans were experts at predicting eclipses, even though they may have not fully understood the reasons for the events. We have no records indicating any understanding of the solar system as rotating planets, or even the round Earth concept. Nevertheless, Moon's cycles and eclipses exerted huge influence on their society in general. Planting, harvesting and social events were aligned with Moon's phases and eclipses, according to rules set by the priests.

We now know that Moon is gradually getting farther away from us by almost two inches per year. It appears that Earth is slowly loosing gravitational control over its celestial companion. Four billion years ago, in Earth's youth, Moon was much closer and its influence much stronger causing volatile weather and huge, 40-60 feet high ocean waves. Moon still has a role in the tides in the oceans, an effect of Moon the Mayans were not aware of.

The Mayans were not only recording Moon's cycles, but those of Venus as well. They called Venus the Great Star, probably due to the fact that Venus was, and still is, the brightest celestial object next to Sun. Their recording of the Venus cycle was also very accurate. They observed its changing from Evening star to Morning star and it is very likely that they knew that they were the same object. After all, the two were never seen together at the same time, and their positions indicated that the Morning Star "visited" with the Sun and reappeared as the Evening Star. We inherited the name of Venus from the Romans, for whom it represented their goddess of beauty, Venus.

Venus also exhibits phases, similar to those of the Moon. In fact, Galileo used the phases of Venus to discredit the Ptolemaic world view of all celestial objects, including Venus, orbiting Earth. There is a full Venus when it is on the opposite side of the Sun from Earth. Then it is of course the farthest away from Earth, hence it is small. Venus shows a Moon like crescent shape when coming around the Sun toward the Earth side. The new Venus, similarly to the new Moon, occurs when the planet is between us and the Sun. This is the time when it is the biggest, however, it is dark.

The ancient Mayan observers were cognizant of the 584 days long Venus cycle. This is the Venus year, the time Venus needs to complete a full circle and arrive at the same position with respect to the Sun, but not necessarily with respect to the sky. Of this cycle, Venus is visible for 263 days as the Evening Star and 263 days as the Morning Star. Venus is not visible for a 50 day long period in between when it is behind the Sun, and for eight days when it is in front.

This cycle is described in a rare surviving Mayan artifact, the so-called Dresden codex, so named after its final location. The codex contains ancient Mayan hieroglyphs describing a sequence of five 584 day Venus cycles. This cycle is of $5 \cdot 584 = 2920$ days, which is eight 365 day long Earth years. Hence Venus appears exactly in the same position in every eight year with respect to the stars, when viewed from the same position on Earth. The Mayans seem to have known that.

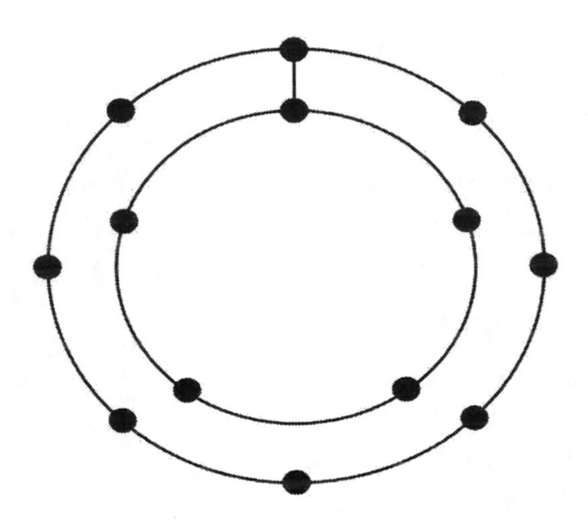

FIGURE 1.2 Venus' cycles

Figure 1.2 depicts this interesting phenomenon. The eight dots on the outer circle represent the location of Venus in eight consecutive Earth years. The five dots on the inner circle represent the Venus years. These positions change by a 72 degree angle from one to the other. Assuming that the top dot is Venus' 2011 position in the sky, it was last seen in 2003 and will be visible again in 2019 in the same position.

The meaning conveyed in the figure is a predecessor of some intriguing phenomenon to be discussed in a later chapter. That is that two different observers (one on Venus and one on Earth) sense different time duration on their respective clocks for the same event. The event here is Venus completing a full circle in the sky arriving at the same location. For an Earth resident that was eight Earth years, while the Venus based observer registered five Venus years. The number 5 appears in various places in the Mayan records, in all likelihood stemming from their Venus observations.

The transits of Venus, the phenomenon equivalent to the solar eclipse by the Moon, was also observed and recorded by the Mayans. Transits occur when an apparently smaller celestial body passes in front of an apparently larger object. The emphasis is on the word apparent, the actual sizes of the objects may be exactly the opposite. The opposite scenario, when an apparently larger object passes in front of an apparently smaller one, is called an occultation.

A transit of Venus occurs when it is passing directly in front of the Sun. Venus is much farther away from Earth than Moon, and the ratio calculation of its size and distance from the Sun does not produce the lucky coincidence we got for the Moon. Hence the Venus transit appears as a small dark spot moving across the face of Sun. The phenomenon usually lasts several hours.

The most intriguing aspect of the transit is its rarity. It occurs in a pattern that is repeated in every 234 years. A 121.5 year long span is followed by two consecutive transits occurring eight years apart. Then another 105.5 year long hiatus follows. The latest and fully documented pair was in December of 1874 and in the same month of 1882. Historical records indicate that Captain Cook even observed the prior one in 1768 on his Tahiti voyage. The next pair, of whose first occurrence already happened, is the 2004, 2012 pair.

After that humankind will have to wait until 2117, 105.5 years after 2012 to see another transit of Venus. Then eight years later, in 2125 there will be another transit, followed by the 121.5 year long hiatus. These again will occur in December as in the 1800s.

The next transit on June 6 in 2012 is an event that is special because it will follow a solar eclipse by the Moon two days earlier. The Mayans' obsession with both Moon and Venus, their eclipses and transits, may have led them to foresee the double event within three days in 2012.

The time difference between 2012 and 3114 BC is 5126 years. Dividing this by 234 yields 21.9, almost 22 full cycles of the Venus transitions. It is plausible that our hypothetical priest started the new calendar cycle after witnessing a similar special double event.

Incidentally the Venus cycle of 8 Earth years is almost exactly 99 lunar cycles, $99 * 29.5 = 2920 = 8*365$, adding to the specialty of the relationship between the two objects, as perceived by the Mayans. The Venus cycle was very important in their lives in connection with larger than everyday events. The installation of rulers, dates of religious rituals, initiation of wars were planned accordingly.

Their god-son Kukulkan, who may have been an actual historical figure, is assumed to be the embodiment of Venus. Kukulkan is known in modern Mexican culture as Quetzalcoatl and the large step pyramid in Chichen-Itza, atop which our ancient priest may have done his observations, was built to honor Kukulkan (or Venus). The pyramid, referred to as El Castillo in modern Mexican history, was excavated in the 1930's and a chapel with a jade throne, presumably that of Kukulkan, was found buried below.

Members of the solar system were not the only celestial objects the Mayans were aware of. They knew about the Star cluster Pleiades or Seven Sisters (after the Greek mythology), since they had a Mayan name for them: Tzabek. The cluster is the closest such object to us that it is visible by the naked eye during winter nights on the Northern Hemisphere.

An interesting, independent source of ancient knowledge about the Pleiades is the so-called Nebra disk. It is a bronze disk of about a foot in diameter with inlaid gold symbols. The disk was found at a site near Nebra in Germany, hence the name, and it was dated to about 1600 BC, making it more than 3500 years old. It was supposedly a very early astronomical clock. The inlaid symbols are interpreted as the Sun, a lunar crescent and a cluster of stars, the latter is assumed to represent Pleiades.

It is very likely that the Mayans also recognized Pleiades as a group, albeit the physical phenomenon governing their togetherness was probably beyond their understanding. It is somewhat still beyond ours, although we now know that they are all moving in the same direction and with the same velocity across the sky.

The Mayans have developed a science in observing the skies and used the observed celestial cycles to measure time. They encountered some difficulties, however, when they contrasted the cycles of Moon to those of the Sun. The lunar cycles between two full Moons (even accounting for their variable nature) did not match with the cycles of the changing seasons, that was in synchrony with the motion of the Sun.

We now know that 12.37 lunar cycles make a full Sun cycle, but the Mayans only recognized that whole number of lunar cycles did not amount to a full Sun cycle. This discrepancy confounded the Mayans and other ancient cultures, ultimately resulting in the abandonment of the lunar time keeping in favor of solar calendars.

2

February: Solar calendars

Observing the full cycle of the Sun led the Mayans to embark upon a solar calendar making activity. Their year was comprised of eighteen months with twenty day duration. This still did not correspond exactly to their yearly Sun cycle observations, therefore 5 day long adjustment periods were added at the end of every year. They started their calendar year with the winter solstice, when the Sun is at its southernmost position.

They also created a mid-range calendar of 52 solar years, the so-called calendar round, that may be viewed as an extension of the solar cycle into a human lifecycle. This calendar was possibly maintained by a high priest throughout his lifetime. They also created a long calendar, the subject of our next chapter, that transcended many lifetimes and even civilizations.

The Mayans were not the only civilization observing and relating to the celestial cycles. The Chinese also had a celestial calendar, dating back as far as 5 millennia. It was a lunisolar calendar, predominantly based on the Moon's cycles. In it, each month started at one hour before midnight of the day of the dark Moon. A year had twelve lunar months which required inserting an additional month every third or fourth year in order to coalesce again with the solar cycle.

The Chinese also had a mid-range calendar with a cycle of 60 years, that was based on the reoccurring conjunction of Jupiter with respect to certain stars in the sky. This particular planet will also be notable in connection with the birth date of Jesus Christ, subject of a later chapter. Whether these cycles served to reconcile their short lunar calendars with the fixed stars in the sky, or something more mythical, is not known.

The Aztecs also had a 365 day long solar calendar, observant of the Sun cycles. There is an original artifact at the National Museum of Anthropology in Mexico City, the famous Aztecs Sun Stone. The calendar appears to have been started with a day referencing a special celestial event, the spring or vernal equinox. That is the day when the length of the day and the night are the same, although we do not know how they measured the duration of the day and night.

The best reconciliation of the lunisolar discrepancy is due to the Babylonians, also several millennia ago. Their long term observation led to the recognition of an intriguing coincidence. They recognized that a lunar cycle or lunar month is about 29.5 days. They noticed that 19 solar years is exactly 235 lunar months. On the other hand that is 19 lunar years plus 7 lunar months ($19 \cdot 12 + 7 = 235$). Hence one could live with a lunar calendar for 19 years and then add 7 lunar months to get back in synchrony with the skies.

That approach, however, would have made them out of synchrony with the seasons for an extended amount of time. Hence they decided to introduce the additional lunar months in installments, a technique that became instrumental in the calendar making and correcting efforts for millennia after that, and even today. They attempted to distribute the 7 months through the 19 years in monthly installments and in regular intervals as much as it was possible.

In a solution reminiscent of our current leap year technique, they added an extra month at the end of the 3rd, 6th, 8th, 11th, 14th, 16th and 19th year of the 19 year cycle. This rather insightful solution resulted in a calendar that was never more than 20 days out of synchrony with the solar cycle. Hence this calendar was also a lunisolar calendar based on lunar months, but attempting to stay in synchrony with the Sun.

The notable role of the number 7 prompted the Babylonians to consider it unlucky. It is believed that the appearance of the week as a time duration was invented by the Babylonians when they declared the seventh day to be the day of rest to avoid any unfortunate events. It is assumed to be the origin of the biblical statements of creation in the book of Genesis: God working on the creation for six days and resting on the seventh. Even in our modern calendars the order of the days is Monday through Sunday, the work week starts on Monday and the rest day is Sunday. In Christian belief the resurrection happened on Sunday.

The Babylonian calendar became very popular in the Mediterranean region in antiquity. Most of the surrounding cultures adopted it. The shortcoming of the calendar was, however, the adjustment years. Their timely execution was dependent on a contiguous rule or governance, and of course an ongoing record keeping. Considering the volatile nature of the region, the rulers and changing allegiances, it was bound to fall out of synchrony.

As we now know and see on Figure 2.1, Earth is orbiting on an elliptical pattern, and the Sun is in one of the focal points of the orbit. Hence when the Earth is on the side of its orbit closer to the Sun, the so-called perihelion, it is moving faster than on the opposite side, the aphelion.

The location of Earth on the left side of the figure represents the winter solstice of December 21st, a time of fear in ancient cultures. It followed a continuing lowering of the Sun's height day by day, and people feared that one day the process would not stop and Sun would not come up at all.

Then of course things would turn around and spring would come with lengthening days and higher locations of the Sun. The bottom location on the figure is the spring equinox of March 20/21st. This is the day of equal length in days and nights.

The Sun would continue to be higher and the days longer until the summer solstice on June 21st. This is the longest daylight day of the year and its position is on the right hand side of the figure. This was a joyous event in many cultures, the so-called mid-summer, celebrated with festivities that are still preserved.

Finally, the cycle concludes by gradually lower Sun

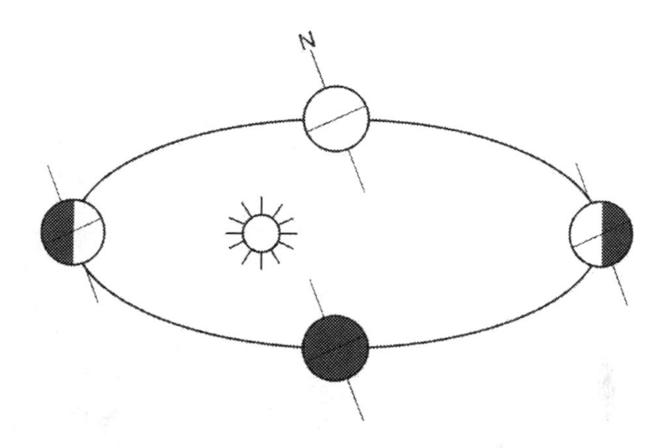

FIGURE 2.1 Solar cycle

positions all the way until the fall equinox on September 21/22nd. It is again the date of equal length of day and night. The lowering heights and shortening days continue until the winter solstice. Apologies to the readers on the Southern hemisphere, for whom the above explanations should be reversed.

Considering the line representing Earth's axis of rotation pointing to north, and comparing the winter and summer solstices, we can see the different sunlight patterns on the two hemispheres. The line representing the equator shows how during the winter in the northern hemisphere, the sun is below, and during the summer above the equator.

While Figure 2.1 implies that Earth is always leaning in one direction, that is far from the truth. The axis of Earth as shown on the figure is not perpendicular to the orbital plane of the Earth. This is in part due to the gravitational forces of the Sun and the Moon. The forces attempt to pull the equatorial bulge of Earth into the plane of Earth's orbit, the plane of the ecliptic. The axis is slowly rotating in the so-called axial precession, that is a rather large, almost 26,000 year long cycle.

Halfway through the precession cycle Earth's axis is going to change such that Northern hemisphere is going to be having winter on the right hand side, i.e. the farther away from Sun. Whether this time will be accompanied by a reversal of the magnetic field of Earth, is subject to much debate. Let it be sufficient to state that there is a much bigger celestial cycle than the yearly solar cycle.

If Earth would have a perfectly circular orbit around the Sun and its axis would have no tilt, Sun would be always at the same point in the sky at the same hour of every day. Since neither of those conditions exist, the location of the Sun in the sky varies from day to day. Recording the location over a year yields a very intriguing pattern of a figure 8. The pattern called analemma is shown on Figure 2.2.

The pattern was probably recognized in the antiquities. One only needs a fixed position retained year

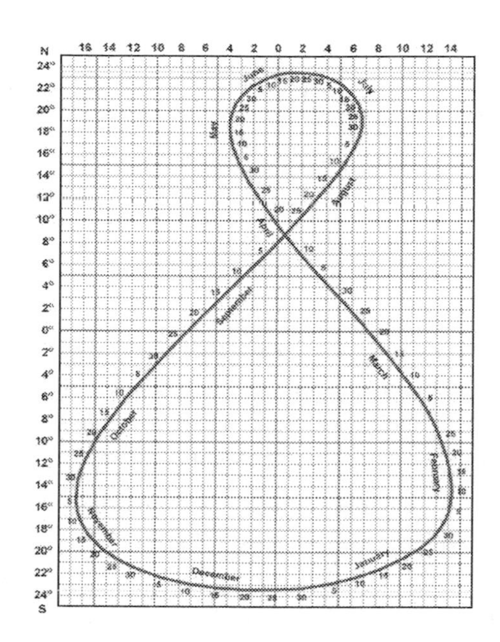

FIGURE 2.2 Analemma

around and a steady orientation toward the sky to observe the pattern. Ancient astronomers very likely had both conditions. It is actually possible to photograph the pattern with modern equipment on a single piece of film by repeatedly exposing it on every day of the year. The camera must be permanently mounted and the pictures taken at the same time on the different dates.

More importantly, the pattern is a very useful time keeping tool. Its vertical axis gives the latitude at which the Sun is directly overhead on a certain day. Its upper maximum is at 23.5 degrees North, which is the tilt of the Earth's axis and the lower minimum is at 23.5 degrees South. The dates are June 21 and December 21, the days of the summer and winter solstices. The latitude locations are the Tropic of Cancer and Capricorn, respectively.

The horizontal axis represents the time difference between the Sun's position and the local time. On the right hand side of the vertical axis the Sun is behind the local time (slower) and on the left hand side it is ahead of the local time (faster). This is due to the fact that the length of the solar day is not constant during the year. When the Earth is on the side closer to the Sun it moves faster, the solar day is shorter, by as much as 16 minutes. On the other side it is longer by about the same amount. The symmetry of the arrangement results in the average value of $24 \cdot 60 = 1440$ minutes per day.

The analemma in essence is a graphical representation of the seasons. Changing seasons were of utmost importance to the Egyptians, a culture highly dependent on the yearly floods of the Nile. They opened their eyes toward the stars in order to produce a repeatable yearly cycle. About five millennia ago they recognized the very regular behavior of a bright star appearing on the early morning horizon, just before sunrise. The star, they called Sopdet and we now call Sirius, appeared very close to the Sun in early September just preceding the yearly flood of Nile. Hence it was called the flood bringer.

We now know that Sirius is actually a binary system of two stars, Sirius A and B. Sirius A is the bright and visible star the Egyptians observed. The companion star B is about the size of Earth, but hundreds of thousand times heavier. Its presence was only deduced from astronomical calculations by the German astronomer Friedrich Bessel in the middle of the 19th century. Based on that, American astronomer Graham Clark actually spotted it about two decades later.

The Egyptians' yearly calendar start was aligned according to the rising of Sirius A and completely abandoned the lunar component of the calendar. They defined artificial but fixed length months of 30 days each and 12 months of those constituted a year. This was only 360 days, just like the Mayans', hence they also introduced a 13th month with five days at the end of each calendar year.

This calendar was installed by the pharaoh Imhotep, sometimes in the 27th century BC. The year had 3 seasons of four months each. The season of the flood was from September through December, according to our current system. The next four months were the season of the vegetation, from January through April. The months from May through August represented the season of the harvest. The calendar was still slightly out of synchrony with the solar cycle, since it was 365 days long exactly.

In 238 BC, Ptolemy III decreed that every fourth year should be 366 days, resulting in a 365.25 days long average over 4 years. The adjustment schedule was kept by the high priests of the church and became known as the Ptolemaic calendar. Ultimately, Julius Caesar, a hero of a later chapter used the Ptolemaic calendar as the basis for bringing the Roman calendar in a better synchrony with the solar year.

So why is the solar year not exactly 360 days long? After all, one full rotation of Earth around the Sun is 360 degrees, and one full rotation of Earth around its own axis is one day. The reason is that Earth simultaneously rotates around its own axis and around the Sun.

Completing one full rotation around its own axis, Earth does not arrive in the same position with respect to the Sun. Each day Earth has to compensate for the fact of being on the orbit around the Sun and rotate a bit more to show the exact same face to the Sun. That is the reason why the number of Earth rotations in a year is not exactly 360.

This argument is true independently of the units of measuring the angle. If Earth were to spin around its own axis faster but on the same orbit around Sun, the number of days in a year would still not be a whole number. That the measure of angle for a full circular motion turned out to be 360 degrees in human history, is simply rooted in the fact that Earth takes approximately that many rotations in a year.

As we saw, the ancient Egyptians viewed their year in a larger frame of reference, that of the stars. But they were not alone, there were other cultures an ocean away doing the same.

3

March: Cosmic calendars

The Mayans opened up their eyes to an even wider cosmic view and created a really long calendar that reaches into our time. It appears that their long calendar was started in 3114 BC. Obviously this date was computed retrospectively since our current time keeping method had not been defined until the middle ages as we'll learn in the following chapter. This is the calendar whose apparent end in 2012 is considered to be the day of Apocalypse by some. As we will see it in a moment, those doomsday predictions are largely unfounded and fueled by the misinterpretation of the long calendar.

Celestial cycles have been observed by the Mayans for millennia and there is no reason to assume that this is their first long calendar. As the new cycles have replaced the old ones, the calendar may also have been re-cycled by the ancient priests. After all the recording material was scarce and very expensive - another component of modern life that may have had thousands of years of history.

There is no question about the fact that some interesting celestial events will take place in 2012. There will be the aforementioned double events involving the Sun within three days, a solar eclipse by the Moon and a Venus transit, in agreement with our Mayan priest's predictions. There is a third, less definitive event of the Sun eclipsing the center of the Milky Way.

The latter is a bit murky because the center of the Milky Way is somewhat of a relative concept, based on our perception that is skewed by the fact that we are in fact part of it. The center is a black hole and we have a hard time comprehending its existence, let alone defining the meaning of that celestial scenario. But, three is a special and magical number, and three major celestial events occurring in a year certainly makes 2012 special, indeed. So the calendar ends, that is a fact. Let us see the calendar in more detail to understand the reasons for its ending.

The Mayan calendar was written with hieroglyphs that were only deciphered in the 19th century. Its cycles were based on a specific, largely base 20 place system. Their digits, separated by a symbol and represented here by a period, designated the multiplier of the appropriate place value. Just like in our decimal (i.e. base 10) system 15 stands for $1 \cdot 10 + 5 \cdot 1$, the value of 2.5 meant 45 days $(2 \cdot 20 + 5 \cdot 1)$ in the Mayan's base 20 system. However, in their desire to reconcile their counting with the years (albeit approximately) the base was switched between the second and third positions to 18 and resumed thereafter.

Hence the place values in their system were: 1; 20; 360; 7,200 and 144,000. These place values had designated names: Kin, Winal, Tun, Katun, Baktun. A date of 1.0.0.0 in their long calendar was 7,200 days, or one Katun. Similarly, the date of 1.0.0.0.0, or one Baktun amounted to 144,000 days, or approximately

394.3 solar years.

To put these numbers in perspective, let us tabulate some years of the current (or as it is called last) cycle of the Mayan long calendar.

TABLE 3.1
Mayan long calendar

Modern date	Mayan date	Elapsed days
08/11/3114 BC	0.0.0.0.0	0
12/06/2720 BC	1.0.0.0.0	144,000
12/07/1143 BC	5.0.0.0.0	720,000
01/01/01 AD	7.17.18.13.1	1,137,141
03/09/830 AD	10.0.0.0.0	1,440,000
09/19/1618 AD	12.0.0.0.0	1,728,000
12/12/2012 AD	13.0.0.0.0	1,872,000
03/09/2410 AD	14.0.0.0.0	2,016,000

The table indicates some specific dates. They may be off by a few days for reasons of various adjustments of the calendars during the millennia, as will be seen in the following chapters. The number of elapsed days from the calendar's start to certain dates shows that the end date can easily be explained.

As the table shows, in late 2012 we will arrive at the conclusion of a great cycle of one Baktun, denoted by 13.0.0.0.0 in their number system. There might be some minor discrepancy in the days and months, since there is no way to know the precise date of the cycle's start, after all those concepts did not exist before. The actual number of days of 1,872,000, puts the start somewhere in August of 3114.

It is quite rational to assume that the Mayans simply started another calendar cycle on the largest place value (Baktun) at that time and that is the reason why it ends in 2012. The fact that it is the end of the 13th great cycle, however, may have some relevance. The Mayan folklore has a story about 13 crystal heads located on Earth. When they were all found and brought together, their combined power would start a new civilization. This would start with the destruction of the old one.

The story does not seem to be completely without merit, after all several crystal skulls have been found in various places on the Earth so far. Skulls are kept at the British Museum, at the Smithsonian Institution and in Paris in a private collection.

One skull, found by the daughter of a British archeologist early in the last century in the Mayan village Lubaantum that is in today's Belize, is a mindboggling artifact. It is an anatomically correct skull with a lower mandible that was created from one crystal and no tool marks of any kind can be found on it. The crystal's chemical composition is similar to crystals only found in Madagascar today and scientists are still debating its possible origin. The technology used to make them appears to be out of the realm of the Earthly technologies of the pre-Mayan times.

Hence today's doomsayers and their gloomy predictions about the end of the world in 2012 may have some justification by assuming that the 13th cycle will be the last great cycle in the Mayan calendar. But there are only 7 skulls found so far, and some of them are suspect of having been made by more recent technology. Therefore, 2012 is likely nothing more than just the end of another great cycle. You dear reader, if you read this later than 2012, are already privy to the knowledge of the truth.

As an additional assurance of our future, note that there are seven more Baktun cycles before we would reach 19.19.19.17.19, the apparent end of their known number system. That would be more than 8 thousand solar years, well in our future. Furthermore, their number system did not even end there.

They had four more, lesser known, higher order (by ratios of 20) place values, named: Piktun, Kalabtun, Kinchiltun and Alautum, enabling one to write numbers as large as 19.19.19.19.19.19.19.19.17.19. This is almost as large as 20 to the 10th power (ignoring the minor glitch of switching to 18 once) and this number is astronomically large, measurable only on a cosmic time scale as it far exceeds even the estimated age of our universe.

Some interpretations of the start of the last calendar assume that the year 3114 BC was the end of the previous world, probably civilization, in Mayan beliefs. It may also have been the date of Kukulcan's founding of their new civilization. Whether a cycle of 5125 years is fitting to some of their historical recollections is not proven, but it is certainly noticeable that their civilization did not last a full great cycle and became extinct much earlier.

There are some records indicating that the Mayans believed in very large cosmic cycles, called the ages of the Sun. The school of thought called Mayan cosmogenesis credits the Mayans with a deep understanding of our cosmic origins. Whether it was self-established or extra-terrestrially influenced is beyond our focus on time, although there is an increasing number of people believing in the ancient astronauts theory, especially with respect to the Mesoamerican cultures, like the Mayans.

The cycles may not indicate the beginning and end of creation, instead they may provide the means to measure the apparent cycles of cultures or civilizations. In the Mayans history there were already four worlds preceding the present one that were destroyed by some events. As we recall, the number 5 held significance to the Mayans from their Venus observations.

The end of the 5th Sun was described by the Mayans as the time when the soul of humans will be released from the material confines (i.e. body) to reach the cosmic heights. This is very similar to other cultures' description of ascending to the heavens and may be the main contributor to the apocalyptic visions.

One thing seems very peculiar: 5 cycles of the Maya long count, $5 \cdot 5125 = 25,625$ is a remarkably good approximation of the length of the precession cycle that takes 25,771 years during which the Earth completes a full wobble against the universe.

We now fully understand the precession phenomenon that is due to the tilting of Earth's axis as discussed in the last chapter. This coincidence is too close to ignore. It is conceivable that the Mayans somehow were aware of the fact that our present North Star, Polaris was once our North Star before. We now know that the Vega was our North Star around 12,000 BC and it will be again in another 14,000 years or so. From 1,500 BC to 500 AD, the Beta star of the Little Dipper constellation "served" as the North Star. Since the Little Dipper is also visible to the naked eye in the night sky and its North Star position overlapped with the Mayan times, they must have noticed the moving of the position of the North against the stars.

The Mayans kept records of their celestial observations though generations and it is not a big leap of faith to assume that they recognized the precession. Such a big time cycle has certain resonance with the theories about the cyclic expansion and collapse of the universe, and renewal by the big explosion again. They were not the only ones who had such a long time horizon, the Hindu time cycles are even longer, in fact astonishingly large.

The Hindu calendars are given in the terms of the life of Brahma. One day in the life of Brahma is 4.32 billion human years. And this is only the daytime, another time span of the same size is Brahma's night. There are 10,000 smaller units, called charanas, of Brahma's day. One of them is a meager 432,000 human years. One month in the life of Brahma is 30 days of his, resulting in 259 billion human years. Twelve of those months constitute a Brahman year, 3.11 trillion years long. Finally Brahma's life span is hundred Brahma years or about 311 trillion human years.

According to Hindu beliefs, at the present we are about halfway through the lifetime of Brahma, in his 50th year. Hence about 155 trillion years are gone by since the incarnation of Brahma. This could be considered as the age of the universe and another 155 trillion years may be before it ends. It appears that we have a long way to go before the universe ends.

Furthermore, Brahma is believed to be reincarnating cyclically, so the universe may not even end then. The scale of this is mindboggling even in a purely arithmetic sense, and the philosophical aspects are far reaching. We will leave those for the reader to ponder further.

The early recognition of our world being a part of great cosmic and celestial cycles, did not alleviate the need for humans to describe the smaller cycles of our lives. This topic was not without its challenges.

4

April: Calendar confusions

Since the Roman empire's territories spread from the deep south in Egypt to far north all the way to the British isles, the agreement between the calendar and the seasonal changes of the Sun was very important. The daylight hours at various latitudes and certain times of the year could vary a lot between the extremities of the empire.

People living close to the Equator and even in Egypt do not see much variation of the daylight hours during the year, but this is not the case for people living on the northern latitudes. On the north, above the 50th latitude like in today's London, during the winter months of the year the daylight hours may be as short as 10 hours or less. On the other hand, during the summer months daylight hours could exceed 18 hours.

The ancient Roman calendar, assumedly established by Romulus, the founder of Rome, had only ten months originally. The months were: Martius, Aprilis, Maius, Iunius, Quintilis, Sextilis, September, October, November, December. It is noticeable that some months follow the Latin name of their numeral order, while some do not. Quintilis, Sextilis, September, October, November and December verbally mean the 5th, 6th,

7th, 8th, 9th and the 10th months. But they are not anymore and we will soon see why.

Martius was named for the God of war, Maius and Iunius were for Roman Goddesses. The meaning of Aprilis may have been designated to honor Aphrodite, the Greek name of Venus. The ten months long calendar with 29 or 30 day months, still trying to follow the lunar cycles, was of course extremely short, less than 300 days. One of the later Roman rulers in the 7th or 8th Century BC added two additional months, Ianuarius and Februarius, to the end of the year and the calendar became 355 days long.

With such a year, the Romans had a noticeable difference when they compared the calendar day dictating the spring or autumn equinox and its day based on measurements that were by that time very accurate. Hence some adjustment was needed.

This was accomplished by a convoluted rule of inserting days into certain years out of every 24 years. These insertions were 22 days each and the insertion years became 377 days long. This rule resulted in an average year of 365 days over a 24 year period. This was clearly an improvement, but it was a system of high complexity.

Therefore it was managed by the Roman magistrate, who dictated the length of each year according to the general principle above. This fact, however, allowed political tampering with the system. The chief magistrate could lengthen the year in which sympathetic politicians were in office and shorten the year of the opponents. This happened on several occasions, sometimes to the extent of a complete misalignment with the solar year. Roman history reports several so-called years of confusion.

Julius Caesar was born in 100 BC (according to our current time counting) into a noble family; his father reached the office of praetor which was the second highest public office. Julius himself engaged in public service early on and also reached the office of praetor by 62 BC. Two years later, in 60 BC he was elected to consul, the highest office. A decade of military activity followed and culminated in his famous crossing of the Rubicon river in 49 BC to end the ongoing civil war in Rome. As a result Julius Caesar became dictator in 48 BC.

By the time of Julius Caesar, the Roman calendar was seriously out of date. Several insertions were missed and celestial events of solstices and equinoxes were falling onto the wrong days or even into the wrong months. Julius Caesar set out to correct the issue, once and forever. While his goal was not fully achieved, his effort was nevertheless very noble, resulting in a system that survived another millennium and a half.

Despite his active involvement in the affairs of the Roman state, Caesar continued his military activities in the two years following his installation as dictator. He successfully conquered Egypt and had a famous love affair with Cleopatra, the Egyptian queen. This, however, may have been more than just a romance of a middle aged man with a young beauty, many years his junior. He became very appreciative of the Egyptian culture whose accomplishments were brought to his attention by the superior intellect of Cleopatra.

It is likely that during that time Caesar became aware of the calendar work of Egyptian astronomers, who had been observing the celestial cycles for thousands of years. They compiled their observations of solstices and equinoxes in detailed records. Caesar supposedly received the method of repairing the Roman calendar from an Egyptian astronomer, named Sosigenes of Alexandria. His recommendation to Caesar was the rather ambitious idea of abolishing the irregular insertion of days and years as they were before, and replacing them with a new system that started with a major adjustment and followed by regular insertions.

Caesar apparently followed the advice and issued his reformed calendar sometimes during the summer of 46 BC, in the month then named Quintilis, the fifth month. He needed to add 67 days to the calendar. The 67 days were necessary because of the missing insertions in prior years. To accommodate the extra 67 days, he cleverly moved the by then last two months following December, that were January and February, to become the first two months, and thereby shifting the others.

Another way to view this is that he just restated the start of year from March 1st, to January 1st. After all, the year is only a cycle whose beginning and end could be arbitrarily chosen. He may have learned that from the Egyptian calendar whose beginning was in September. This change is the reason why the months' names do not reflect their numerical order anymore.

But this was not enough for 67 days, so he added to the lengths of the various months, hence some became 30, some 31 days long, as:

$$31, 30, 31, 30, 31, 30, 31, 30, 31, 30, 31, 30.$$

This year was a regular alternating sequence of long and short months, but now the year was 366 days long. In order to correct this, he proposed that February be 30 days long only once every 4th year. The other three years it was to be 29 days, resulting in a 365 and a quarter day average over every four year period, just like in the Ptolemaic calendar. The fourth year, when February is longer, is what we call the leap year today.

This naming convention is worthy of an explanation. What happens in that year is that if February 28th is a Monday in a leap year, the same date of February 28th will be a Wednesday in the following year. This is due to the extra day that results in the Tuesday being "leaped over" in the next year. A simple rule that cannot be ruined, so it seemed.

There was a mistake, however, in implementation. The extra day was added every three years, instead of every fourth, and the calendar got out of synchrony again. The reason for the mistake is unclear, but records show that 45 BC was declared a leap year, possibly to account for the missed 46, that was supposed to be the first leap year. The next legitimate leap year was 42 BC and this three years difference created an unfortunate precedent. The leap date insertion continued every third year for three decades before it was corrected.

The Senate later honored Julius Caesar by renaming the original 5th month Quintilis to Julius, in appreciation of his contribution to the Republic. Julius was assassinated on March 15, 44 by his adopted son, Brutus. His death resulted in a huge societal turmoil that lasted more than a decade. The state of Roman affairs did not really stabilize until Augustus, the first ruler of the new Roman Empire, as opposed to Republic. Augustus, a nephew of Julius Caesar ruled from 27 BC to 14 AD, thus transcending the time of the birth of Jesus.

He continued Julius' work on the calendar. It is notable that two relatives contributed to the correction of the Roman calendar. He corrected the leap year spacing error in 8 BC, abolished the three year leap cycle and prescribed it to be four.

According to rumors, Augustus himself changed the eight month, originally called Sextilis, to Augustus, in a bit of a self-promotion. Some other rulers with monumental egos, also made attempts at renaming certain months, but they did not stick. Nero, for instance decreed the new name of Aprilis as Neroneus, but after his bloody reign, nobody wanted to remember him every year.

Furthermore, some anecdotal records indicate that Augustus detested the fact that the month of Julius was 31 days long and Augustus only 30 days. Therefore he "stole" one day from February to make it the 28/29 days as we know it today. He added it to Augustus, to make it also 31. Then in order to still maintain a long/short sequence, he changed the rest accordingly to:

$$31, 28/29, 31, 30, 31, 30, 31, 31, 30, 31, 30, 31.$$

This irregular long/short sequence of calendar days still causes problems to countless generations of school children having to learn the sequence of the months by engaging in the hand game of counting knuckles and valleys.

The most common method of counting the years by the Romans was by counting them from the inauguration of the new pair of consuls. Beginning in 153 BC (per our present counting) they usually started the office on New Year's day. The years counted in this fashion were called the consular years. This counting was cyclical, the year count was set back to zero at the beginning of every new consular period. Well, maybe to one, the zero as an algebraic entity was not widely accepted in Europe yet.

There was yet another system, in which the years were counted from the founding of Rome. This was, however, not used by the common folk in everyday circumstances, mainly because its backward time horizon was just excessively long (although such a fact did not bother the ancient Mayans at all). This was used mainly for official recordkeeping by ancient historians.

To make the interpretation of the Roman calendar record even more convoluted, years were sometimes counted starting from the beginning of the reign of an emperor. The years in this counting system were called the regal years.

Both consular and regal year counting were in use all the way past the middle of the first millennium. Their usage attests to the difficulties of humankind to establish a uniform time scale even within one particular empire, let alone world-wide.

Finally, in 525 AD (per our current count) the philosopher Dionysius proposed to align the year counting to the birth date of Jesus. His system of Anno Domini (AD or the year of the Lord) soon permeated the whole Christian world. We now denote the dates before the first year AD by BC for Before Christ.

There are conflicting stories about the correctness of the birth date. According to one school of thought, he was referring to the date of Jesus' conception not the actual birth date, but the date would be approximately correct. There is another camp of historians who propose that Dionysius actually made an error of several years.

They base their hypothesis on the motion of Jupiter, considered to be the factual star of Bethlehem. Most of the time the outer planets, such as Jupiter appear to be moving west to east, since we on the Earth are on the inner circle. But at the proximity of the point when Earth "overtakes" an outer planet, a strange phenomenon called retrograde motion occurs. During this, the outer planet appears to be moving opposite to the normal direction, i.e. from east to west.

According to astronomical retro-calculations, starting from August 23rd until December 19th in 6 BC per our current count, Jupiter was in its retrograde motion. This could have been the star the three biblical kings followed west to Bethlehem.

Another historical fact supporting a date other than the one chosen by Dionysius is that according to some biblical records Jesus was born under the reign of Herod the Great. Herod, however, died in 4 BC, a well documented fact. Hence it appears more likely that Dionysius was off by about 6 years, but it does not seem to matter for our current state of calendar keeping.

There are some childish efforts to make this politically correct in recent years by changing those names to BCE and CE, for Before Common Era and Common Era. They don't seem to be succeeding as people aren't really bothered by the old notation used for one and a half millennium, regardless of whether they are Christians or not.

Even with the adjustment day and the leap year sequence, the Roman calendar accumulated noticeable error through the first millennium. The Julian calendar year (in agreement with its origin, the Ptolemaic) was exactly 365 days and 6 hours long. The solar year, as we now know very accurately is 365 days, 5 hours, 48 minutes and 46 seconds. The discrepancy is 11 minutes and 14 seconds, or 674 seconds. Since a day consists of 86,400 seconds, the 674 seconds yearly error of the Julian calendar amounts to a full day error in every 128.19 years.

While that does not seem to be of big consequence,

it is significant in historical perspective. The Julian calendar fell into considerable discrepancy regarding certain calendar days of societal importance, like the beginning of Easter. That date was officially defined in a specific relationship with the spring equinox of March 21. By the middle of the second millennium, the Julian calendar was about ten days ahead of the celestial schedule and this required another correction.

Pope Gregory XIII was born in 1502 as Ugo Boncompagni and died in 1585. He studied law in his youth and started a legal career that included teaching law to some of the high members of the clergy. He lived a lay early life, even had a son out of marriage, but later was ordained. By his late thirties he was working for various papal offices, including serving as a legate to Spain. He became a cardinal in 1565 and was elected to become Pope in 1572.

His legal background likely contributed to his reform minded reign as Pope; he set out to introduce significant changes to the Catholic church. Those included positive rules that required that bishops reside in the district they oversee, and some negative actions such as maintaining a list of forbidden books.

His most important contribution as a reformer was, however, that he organized a committee to work on mending the Julian calendar. His interest was naturally closely related to the church, specifically to that of the date of Easter.

The committee included several disagreeing members, among them Aloysius Lilius, a medical doctor, whose idea prevailed in the end. He proposed to use a length of the year published in the middle of the 13th century in the so-called Alfonsine tables. They were named after the Spanish king Alfonse X, who commissioned the publication.

The Alfonsine tables listed a mean calendar year of 365 days, 5 hours, 49 minutes and 16 seconds. That was still 30 seconds longer than the true calendar year of 365 days, 5 hours, 48 minutes and 46 seconds. This was, however, much better than the Julian difference of 674 seconds.

Lilius' main contribution was to find the way to bring the Julian calendar into synchrony, a difficult thing to accomplish. The difference between the two was 644 seconds. Dividing 86,400, the number of seconds in a day by 644 yielded 134.16. This meant that the Julian and Alfonsine count was off by a day in every 134 years.

Some other members of the committee proposed simply dropping one day in every 134 years, certainly a conceptually simple, but administratively cumbersome if not impossible solution. Lilius, however, recognized that 3*134 is 402, or almost 400 years. This led him to the easily manageable rule of ignoring three leap years in every 400 years which is still the basis of the current leap year rule.

Let us now consider Gregory's main concern, the date of Easter. This was a confusing topic for hundreds of years and it is still somewhat clouded today. The various fragments of the Catholic church celebrated Easter on different days. The Catholic church at the time of Gregory defined Easter as the Sunday after the first full moon that follows the spring equinox. The date of the spring equinox was March 21st, hence the definition could place Easter anywhere between March 22nd and April 25th.

This variable date was cause for confusion throughout the ages, so much so that even luminaries of science, like Gauss, spent time devising computational algorithms. The fact that one can do so instead of waiting for the actual full Moon to occur is obvious, since we now know that the lunar cycles are predictable. Let us now follow Gauss' method and calculate the date of Easter for the year 2011.

The calculation starts with dividing the year with 19 and retaining the remainder. The number 19 is important because it is the length of the long lunar cycle, discussed in detail earlier. Hence for the year of 2011 the remainder is 16, because:

$$2011/19 = 105 \cdot 19 + 16.$$

The next step is to multiply the remainder by 11 and subtract this number from 225. This in our case yields $225 - 11 \cdot 16 = 49$. If this number is greater than 50, subtract 30 until the number is below 51. If it is greater than 48, subtract 1. This is our case, and we will save this number as $a = 48$.

Now consider the year again and divide it by 4, but retain the whole number part of the division. Since $2011/4 = 502 \cdot 4 + 3$, we get 502. We will save this number as $b = 502$. Finally we compute a third number by adding:

$$year + a + b + 1 = 2011 + 48 + 502 + 1 = 2562.$$

Dividing this number by 7 and retaining the remainder again, for our case: $2562/7 = 366 \cdot 7 + 0$. We now save this remainder as $c = 0$, Subtracting this number from number a from above and adding 7 we will obtain d, the Easter date. For our case:

$$d = a - c + 7 = 48 - 0 + 7 = 55.$$

If d is greater than 31, the date is April d-31. For 2011 the date of Easter is d=55-31=24, April 24th, a rather late Easter. If d is less than 31, the date of Easter is on the d-th day of March.

The calculation is applicable to the Gregorian calendar and valid only until the end of the 21st century. Minor adjustments to the arithmetic will carry the calculation further and there are methods that are valid forever, a rather long time. The more forward looking the prediction needs to be, the more complex the calculation process becomes.

Due to Easter's religious importance, establishing the precise date of the spring equinox was crucial in the calendar. By 1582, the accumulated error in the Julian calendar showed March 11 as the day of the spring equinox that was supposed to be March 21. Clearly there was a discrepancy that needed to be corrected.

The Papal decree by Gregory in 1582 ordered the day after October 4 not to be October 5, but the 15th. This change, however, only corrected the existing discrepancy between the Julian calendar and the celestial cycles. Without any further change the new calendar would also gradually become misaligned in the future.

Hence the reform also required the modification of the Julian leap years, as proposed by Lilius. The redefined rule was that the leap years (that are in general in each fourth year) are skipped when the year is a full hundred but not divisible by 400. Hence 2000 was a leap year, while 1900 was not and 2100 will not be.

With this new definition, the average Julian year of 365.25 days was reduced to 365.2425 days. Precisely, the Gregorian year became 365 days, 5 hours, 49 minutes and 12 seconds long. The deviation from the solar year became only 26 seconds. This, still approximate year length decreased the accumulated discrepancy by several orders of magnitude. No adjustment was going to be required for another 3323 calendar years. However, only 500 years later another problem surfaced.

5

May: Missing years

The adoption of the Gregorian calendar was far from easy because the Papal decree held no authority outside of the Catholic Church and the states adhering to the religion. These were Italy, Spain, Portugal and France; most adopting it immediately during 1582. The Protestant states, however, resisted as they considered this as an effort of the Catholic church to bring them into its fold again. They had a civil calendar that was the rule of their land.

Ultimately the Protestant countries also adopted the change, albeit much later. By then the change had to be adjusted to the increased time discrepancy and some difficulties arose. An interesting example is Sweden, that decided on the adoption in 1700 and by that time skipping 11 days was needed. Instead of skipping eleven consecutive days, however, the throne decided to skip 11 consecutive leap days, starting from February 29, 1700. This course of action meant that the calendar would adhere to neither the Julian, nor the Gregorian calendar until 1740.

Apparently governments were not very effective in those days either; as fate would bring the Swedish immediately forgot to skip the leap day in 1704. Furthermore they even forgot to skip the leap day in 1708, so instead of being only 8 days behind the Gregorian calendar, they were still behind by 10. Since the original gradual plan was thus rendered useless, Sweden was forced to go back to the Julian calendar. This was done by introducing an unprecedented February 30 in the leap year of 1712. So Sweden was back to square one and entered into the annals of time trivia.

The 12 year long exercise in futility was so discouraging that it took Sweden almost another half a century to correct it. Finally in 1753 the feat was accomplished by jumping from Wednesday February 17th to Thursday March 1st. This action demonstrates the difficulties of changing calendar time, in order to maintain continuity of the weekdays a specific date must be chosen.

If the Protestants were so much behind in adopting the Gregorian calendar, the Eastern Orthodox Christian states were even more so. In a sense they never really adopted it. The Eastern Church performed a correction of the Julian system less than a hundred years ago, in 1923. They skipped 13 days at the time and declared Wednesday, February 15, 1923 to be followed by Thursday, March 1, 1923. The Eastern Church also adopted the leap year rule, but calls this a "revised Julian calendar" (as opposed to a Gregorian calendar) that in essence will be in synchrony with the Gregorian for another 800 years.

But, even within the Orthodox Christians there is no agreement. By today, Greece, Romania, Bulgaria and Cyprus have adopted the revised calendar, but Serbia, Macedonia and Georgia have not. The latter countries celebrate Christmas on December 25th in the Julian calendar, that is January 7th in the Gregorian calendar. Time and calendar are apparently sensitive components of church philosophy.

Besides the checkered history of its adoption; the Gregorian reform also brought some controversy and even fueled some conspiracy theories. The fact that the average Julian year was slightly longer than necessary meant that it was going ahead faster than the celestial events. March 11 in 1582 being already the celestial spring equinox, that was supposed to be on March 21, demonstrates that. The calendar date of a celestial event in the Julian calendar is behind the actual celestial event, because its calendar unit is longer.

Having digested this subtlety, we can understand the fact that Pope Gregory "adding" ten days or "omitting" ten days from the Julian calendar to synchronize it with the celestial events describe the same effect.

It was, however, not noticed for centuries that the ten days adjustment of Pope Gregory was not commensurate with the amount of time between his and Julius' correction. The elapsed time between the two adjustments was 1627 years, 1582 AD + 45 BC. According to the Julian calendar's length, the discrepancy was about one day in every 128.18 years. But 1627 years divided by 128.18 yields 12.69 days!

Something is obviously incorrect here. According to this simple and uncontestable calculation Pope Gregory should have added (or omitted if one prefers this way of looking at it) almost three more days. The ten days, using the 128.18 year per day number, results in only 1282 years.

The difference between 1627 and 1282 is 345 years. Accounting for the actual months of the adjustment years, the resulting difference is considered to be some 300 years. Gregory's adjustment is apparently short by those 300 years and they seem to be missing years from history. The ensuing controversy is interesting and worthy of a brief review.

There are some historical and even archeological facts that seem to prove the missing years hypothesis. The historical facts hinge on the so-called "dark ages". The years between the 600s and 900s are called such because there are barely any credible references actually written in that time period. References written later, but describing events and lives of historical figures during those years, are considered to be forgeries by some.

There are several unexplainable historical references of certain events allocated to distinctly different dates of about 300 years apart. For example, certain documents recorded the famous burning down of the library of Alexandria in either 47 BC or 272 AD. The destruction of the Basilica Julia in Rome is also reported to have been in 12 BC and in 283 AD.

It is certainly possible that these two catastrophic events could have happened twice in the span of centuries. It is, however, implausible that a notable event of a particular person's life would happen on two different dates three hundred years apart. Yet, the well recorded event of the mother of Emperor Constantine receiving the Holy Cross in Jerusalem is dated to 325 AD and also to 629 AD.

Then there are recorded events attributed to both Constantine in around 310 AD and to Heraclius in about 610 AD. That is again highly unlikely and we must again suspect the date keeping since the number of years between the dates is in the neighborhood of the missing 300 years.

Finally, there are also solar eclipses in certain observation records that seem to have occurred about three hundred years before their retro-calculated dates by modern science. The observers of solar events are usually meticulous scientists whose records are trustworthy, but here contradict modern science. The evidence of some timekeeping error keeps piling up.

A very intriguing historical gap exists between certain European ruling families leading to a far reaching hypothesis that Charlemagne was not a living person, but an artifact of a large scale history falsification. This hypothesis was put forth by the German historian Heribert Illig, and some archeological facts seemingly support his proposition.

The cathedral of Aachen in Germany, the purported resting place of Charlemagne has yet to yield the remnants of the ruler, despite extensive efforts with modern tools. There are also difficulties in confirming the date of the construction of the cathedral. Various scientific dating techniques, some of those will be discussed in the chapter dealing with measuring time, produce an age that is inconsistent with the age from the recorded time of its construction.

Another spectacular example of historical mysteries that could be explained by the missing years is the story of the descendants of Attila, the Hun. He roamed a vast territory in middle Europe, centered about the Danube river in the heart of today's Hungary. He lived from 406 to 453 and was famous for his horse-back raids into the surrounding countries. His marauding troops were the curse of Europe at the time.

His escapades spread as far as Toulouse and Orleans in France to the west, to Constantinople and Rome to the south. After a long campaign through Italy, he returned to his home by the Danube in 452. The circumstances of his death are unclear. According to some sources he was killed by his young wife, the beautiful Ildikó (a popular Hungarian name for women today, as is Attila for men). Other records indicate an unexplained bleeding sickness.

Now comes the connection with the missing years. Some Hungarian chronicles state that Álmos, the historically proven leader of the Hungarians when they permanently settled the Danube elbow region, today's Hungary, was Attila's great grandson. According to written records Álmos appeared on the scene in the first half of the 800s. His own grandson, Árpád, who lived between 845 and 907, founded the formal Hungarian state in 896. That date is proven by historical records.

There is, however, a conflict between the time period and the five generations between Attila and Árpád.

Five generations usually amount to about 150 years, assuming a generation of about 30 years. The elapsed time between Attila's death in 453 and the founding of the Hungarian state in 896 is almost 450 years. That is about 300 years more than the five generations would warrant. That difference sounds awfully similar to the 300 year adjustment error by Pope Gregory.

If one introduces the missing years, that would place the year of the Hungarian state's founding to around 600. Then the succession from Attila to Álmos, and to Árpád in five generations makes the timeline plausible. Of course one can always question the records of the descendants of Attila and argue that there were more generations between them. Since this is a very credible possibility, we will leave this line of investigation to the historians and return to the time aspects.

That there is a time adjustment error seems unquestionable. One must assume that Pope Gregory acted in good faith and in view of the indisputable difference between the calendar and celestial dates. Hence the years by his time must have been denoted as the 1500s. The proposed solutions for this mystery range from astronomical events through outlandish sounding conspiracy theories to the mundane.

There is a school of thought entertaining the idea that sometimes during the dark ages Earth received some kind of an extra-terrestrial impact altering its orbital behavior. This is not impossible since the prevailing view of astronomers acknowledge prior catastrophic impacts during Earth's history. Whether such could have caused the decrease in the discrepancy between the Julian calendar and the celestial events, resulting in the smaller than necessary correction by Gregory, is questionable.

Professor Illig's followers see the trails of a deliberate time falsification by Pope Sylvester II who ascended to the papal throne in 999 according to our current year counting. We have to take note, that the AD counting of years was not widely accepted throughout Europe yet, since it was only proposed in 525 AD by aforementioned Dionysius.

If we take away 300 years from 999, Pope Sylvester II may have been starting his reign only around 699 and amidst a mixed year counting system between the countries. It is conjectured that in his egotistical desire to be the Pope occupying St. Peter's throne at the historical millennial time, he declared that year to be 999 and initiated the new millennium.

It is also possible that he did that only retroactively in 701 AD (accounting for the missing years) and declared it to be 1001 AD. His reason could have been to alleviate fears of the end of the world by the 1000 year anniversary of Jesus' birth. In either case, he was in the position to order such a change and the scenario cannot be dismissed as completely implausible.

Such a radical time change declaration could have not been done at that time without the consent of the German half of the empire. It is proposed that Otto III, ruler of the German territories was Pope Sylvester's conspiratorial ally. This track is now getting too controversial and we'll abandon it cautiously.

A more plausible solution for the mystery of the missing years, however, may lie in the topic of the various year counting systems mentioned before. The Byzantine empire by the time was as powerful, if not more, than the Roman. The Byzantines did not adopt the Roman year counting system of AD, but followed a regal year counting system of their own.

The start of that system was in 311 BC at the beginning of the reign of Seleucid I in Babylon. He was an officer of Alexander the Great and ruled the remnants of Alexander's empire in the Babylon territory. Hence his calendar was sometimes also called the Seleucid calendar. Counting years from his reign provides a discrepancy almost exactly equal to the number of the missing years. So we arrived at a possibly rather mundane explanation.

The Seleucid year counting was used all the way to the end of the 6th century according to some records found in todays Syria. It was also used in Yemen until modern times. Somewhere in the first part of the 600s, there may have been a reconciliation between the Byzantine and Roman year counting system by the latter adopting the earlier, jumping from early 600s to early 900s. The strong ruler of Constantinople at the time, Constantine VII may have had the clout to force the Pope to agree to that.

This scenario is even more credible if one considers the fact that the years, even when they appeared in documents, were not marked by the designation of the system, i.e. after Seleucid or after Christ. Byzantine chronicles strictly denoted years by numbers, without a reference date.

The missing years, however, could also be explained another way. We have almost three days in question. Recall the Romans' error of assigning leap years after Julius' death in every three years until Augustus corrected it. Hence the leap year sequence from Julius to Augustus was:

$$45, 42, 39, 36, 33, 30, 27, 24, 21, 18, 15, 12, 9.$$

This sequence gave 13 leap days in 36 years, as opposed to the required 10 only. Augustus corrected it in 8 BC and the following leap years became 4 BC, 0 AD and 4 AD as leap years, and so on until today.

If the year 0 AD is confusing, it must be noted that it was not called as such at the time of its happening. The actual introduction of the Christian counting was done in the 6th century as we noted above. Had the years been counted from the time of Julius' reign (48 BC) for example, those dates would simply be 44 AJ, 48 AJ, and 52 AJ, using the just invented notation of Anno Julii, or the year of Julius.

What happened is that the Romans took away three extra leap days unnecessarily during the 45 BC to 9 BC period. These three days are exactly the same three days Gregory did not have to take away. Had the Julian error not occurred, the date of spring equinox in 1582 would have been only the 8th of March. Gregory would have then had to skip thirteen days, and that fits the time discrepancy between 45 BC and 1582 AD.

This more rational explanation is also supported by celestial facts. A simple adjustment of a certain number of years would have put certain celestial cycles, well observed for centuries if not millennia, out of synchrony with the calendar.

The Moon does have some regular and well known long cycles. We mentioned earlier the Babylonians recognition of the 19 year long lunar cycle in the construction of their calendar. That was the basis of synchronizing the lunar and solar cycles. The long lunar cycle also brings back the same lunar eclipse patterns in a regular fashion and since they were very special events, they were well observed. By the medieval times most Mediterranean nations kept meticulous track of those events.

Hence, the calendar cannot be adjusted, for example, in the 5th year of the 19 year long lunar cycle by 300 years because that would put the new date into another place in the lunar cycle. Discarding the possibility that a change of an exact multiple of the long Moon cycle was executed, one can conclude that any arbitrarily changing of the years would noticeably violate those observations.

There are no apparent discrepancies with the celestial events and since historians are fiercely debating the conspiracy aspects of the timeline, the most likely explanation for the 300 years conundrum is that the shortened adjustment was due to the incorrectly applied leap days by the Romans.

The truth of this explanation would not eliminate

the age discrepancy of the Aachen cathedral, but would reconcile the timeline of the Hungarian history and even let Charlemagne rest in peace as a person who once really lived. So we will leave this mystery with that most benign explanation.

Before continuing, however, we must make note of the fact that when we quoted year lengths in days, hours, minutes and seconds, those numbers were computed in retrospect. There was certainly no equipment measuring minutes and seconds in the Babylonian or ancient Egyptian times. The concepts were not even invented yet. The subdivision of the smallest celestial time unit, the day, was the next challenge of humankind.

6

June: Measuring time

The Egyptians had sun-dial devices as early as 1500 BC. They were sticks stuck into the ground and the time measurement was done by observing the shadow of the stick. The stick cast a long shadow at sunrise, toward west. As the Sun was rising, the shadow came closer to the stick, until by noon it was close to the foot of the stick. On certain locations, like on the Tropic of Cancer the shadow actually disappeared at noon, when the Sun was exactly overhead. Then the afternoon Sun turned the shadow toward east and gradually increased its length. By sunset, the long shadow was pointing to the east, reversing the progress from the morning.

While it was possible to rely on the Sun to measure the passage of time during the day via sundials, the continuous nature of Sun's motion made the quantitative measurement difficult. Where was 9 am, or 3 pm on the sundial shadow? Furthermore, why did the day become 24 hours long, why not 10 or 20?

According to the current wisdom, the use of 12 over 10 was due the Egyptians, who were very adept in counting fractions and working with them in various algebraic computations. Their preference of 12 over 10 may have been because 12 has four integer factors: 2,

3, 4, and 6, while 10 has only two: 2 and 5.

Their algebraic preference may also have been reinforced by celestial events. Some records indicate that in ancient Egypt in the summer 12 stars arose during the night. Their rising was, of course, not exactly hourly per our current clocks, but their number was the defining fact. In any case, the hourly timekeeping became standard only with the advance of mechanical clocks, developed much later.

The sundials also had a problem on overcast days and during the night, due to lack of sunshine. The next advancement in time measuring devices were the water clocks, also invented by the ancient Egyptians. Their big advantage over the sun-dials was their ability to keep time during the night also. Their disadvantage was the need to regularly replenish the water.

To alleviate the need for the regular water refill, the water clocks were later connected with a continuous water source. To quantify the now continuous time flow, various mechanical instruments were invented. The most common versions relied on a floating device that had some stylus attached. The stylus pointed to a scale with markings for certain time durations, like hours. The continuously flowing water filled a barrel and the rising water level lifted the flotation device, most likely made out of cork. It was still necessary to reset the clock by emptying the barrel when it became full, but only once a day.

Water devices were not portable, however, therefore self-contained sand based hourglass timekeeping devices dominated the middle ages. This was especially true to the most common means of travel at the time, ships. Ferdinand Magellan, during his historic circumnavigation of the Earth, used 18 hourglasses to measure the time of daylight. The first glass was turned on at sunrise, then when it wound down, the second was turned upside down and so on. The number of wound down glasses showed the number of hours elapsed since sunrise.

Humankind continued to invent new clocks of various kinds. The name of the device probably comes from the Latin word clocca meaning bell. Since on ships the passing of time was usually marked by ringing a bell every time an hourglass wound down, this led to the name. Then came the more advanced devices in church spires and the accompanying bell ring at every hour, or later even a chime at the quarters. Those were driven by gravity via weights attached to a rotating cylinder as they still are today.

The ongoing quest for smaller and smaller measuring tools led to the invention of more and more complex time devices. The spring powered watch was probably invented in Germany in the early 1500s. The first versions were rather inaccurate, since the spring's winding down speed was dependent on its coiling tightness. A tightly coiled spring at the start is unwinding faster than toward the end of its cycle.

These were also the first portable watches. At the beginning people were wearing them on a chain around their neck and it was a rather prestigious symbol of status to own one. They were also serving as jewelry.

Diamond or gem stones were embedded into the face that was not yet covered by glass as we know it today. There was a cover, sometimes solid on a hinge, and sometimes a grill work enabling to see the time without opening it.

The second half of the 16th century brought the minute hand by Swiss clockmakers, whose mastery of the trade is still admired today. Rumor has it that the very first clock with a minute hand was built for Tycho Brahe, the famous Danish astronomer, to aid his celestial observations.

The portable watches emerged in the pocket form early in the 17th century. Allegedly the trend was created by the new fashion of waistcoats. Whether it is true or not is not clear, but it is a fact that ladies were still wearing their portable watches on a chain as a pendant well into the 19th century. It is also believed that the originally square portable watches evolved into a round shape to ease removing them from a pocket. But there was much more to come.

Humans felt that the heart beats many times during a minute and considered this to be a reasonable smaller unit to measure. That unit became the second. The "second" name is actually the remnant of the original designation of "second minute". Since it was a bit difficult to detect one single heartbeat, and heartbeat was subjective, a clock measuring that small time duration in an objective way was desirable. This was the pendulum clock.

The idea of a pendulum clock was originated by

Galileo Galilei, a frequent hero of our story. Galileo recognized that the frequency of the motion of a suspended pendulum is independent from the mass of the pendulum, but related to the length of the rope. Rumor has it that he realized this fact by using his heartbeat to measure the period of the swinging lanterns hanging from the ceiling of his church. He actually designed a clock based on a pendulum, but he was not able to find a craftsman to produce it.

Christian Huygens, the famous Dutch physicist completed Galileo's work and created the first working pendulum clock. This enabled the measurement of fixed seconds and Huygens defined the 3,600 seconds long hour duration. The origins of the 60 base for time keeping is attributed to Babylonian place system of the same base. Huygens' original pendulum clock, built by a Leiden clock maker in the 1650s is still exhibited in a museum there.

Huygens was born in Hague in 1629 and also died in his city of birth in 1695 after a spectacularly successful scientific life. He studied at the University of Leiden, still a scientific stronghold today. His main interest was physics and his scientific accomplishments are impressive. His theory of the light waves is his most important scientific legacy.

He also noted that the pendulum's period is a function of the location of the weight and not its mass, leading to the adjustability of the clock. Still today, the weight at the end of the pendulum (the bob) is on a thread, hence the length of the pendulum and its accuracy may be adjusted.

Furthermore, he recognized the fact that the pendulum's swings are not isochronous, or executing in the same time duration. There is certain decay in the period of the pendulum swing that depends on the hinge mechanism and its friction. Huygens devised a mechanism that advanced a gear at every swing and at the same time exerted a nudging force on the pendulum to retain its cycle time. This was the final component to make the pendulum a reliable time measuring device, truly a clock.

He intended to develop naval clocks to be used on the high seas. Unfortunately the motion of ships rendered the pendulum clock useless as it was unable to operate when not vertical. No matter how ingenious his hinge mechanism was, the physics of the pendulum was incompatible with ships.

It was known to the Mediterranean sea-farers centuries BC that the longitude location is directly related to time. Since the Earth rotates 360 degrees in 24 hours, it is a simple division to figure out that it rotates 15 degrees per hour. As a consequence, when a ship is traveling on the sea, for every 15 degree change in longitude the local time varies by an hour. The ship traveling east looses one hour of the local time, i.e. the clock moves ahead.

This is manifested today, when one is taking an airplane to the east. The day is very short when flying non-stop from Los Angeles to Europe. Leaving in the morning gets the plane into Europe the following morning. The explanation is the 9 or 10 hours extra time lost due to the longitude difference.

Conversely, a plane traveling west benefits from the direction and gains an hour in every 15 degrees traveled westward. Leaving Europe in the morning puts us into the United States by noon, despite the fact that the flight time is 10 or eleven hours. This gets a bit more confusing when one travels between the US and Asia. Strange things happen when one crosses the international date line as we will see soon.

Huygens was familiar with the issues of naval timekeeping and the so-called longitude problem. The problem was keeping a reference time when traveling on the open ocean. While the local time was measurable based on the Sun, in order to find out the longitude location the time at a reference point was also needed. Simply a device was needed on board that was able to reliably keep the time at the home port.

In the 17th century, prior to time zones and airplane travel, the problem was difficult. King Charles II in 1675 issued a challenge to the scientists at the Royal Observatory to solve it. Despite almost a half a century of research (including that of Huygens) and numerous failed solutions, the problem remained unresolved. The British admiralty offered a reward of 20,000 pounds sterling in 1715 to solve the problem with an accuracy of half a degree in longitude, or two minutes of time.

John Harrison was born in Yorkshire in 1693 and followed the footsteps of his father by becoming a carpenter. He received no formal science education but still, became very interested in time keeping mechanisms already as a teenager. He built a wooden clock true to his trade in his early 20s, and developed a reputable clock business with his younger brother strictly using wood in his clocks.

Later he became engrossed with the longitude problem. The solution of the problem was a time keeping device that was portable and not subject to the physical effects acting on the ship in the seas. He built a portable version of his wooden clock in the early 1730s with moving parts that were controlled by springs and screws, hence completely independent of gravity. This was an important consideration due to the motion of the ships.

Harrison's chronometer was tested on a voyage to Lisbon aboard a ship named Centurion. It performed very well, however, was deemed not to reach the accuracy requirement of the Admiralty. This evaluation is highly suspect in retrospect, after all, how could the Admiralty measure the accuracy when there was no such device available yet?

Nevertheless, Harrison took on the challenge of further improving his device. He spent almost three more decades and redesigned his device three times. Another test was sanctioned by the Admiralty in 1764 and this time Harrison's son, William accompanied the clock on a trip to Barbados. The fourth incarnation of Harrison's design was on board and was accurate to an error of less than 40 seconds over a journey of 47 days. It was considered a great success by all, but for the Admiralty.

They declared the clock to be a lucky, one of a kind piece of work and refused to issue the price. They demanded from Harrison to turn over the design to the Admiralty for 10,000 pound sterling. The second half of the price would be paid only when other devices built from the design proved to be accurate on 30 mile long longitudinal time trials.

What would follow is a long story, worthy of another book on its own right. It shows the pettiness of the establishment regarding the commoner who answered their challenge. Despite several copies made, independently tested and proven to be three times as accurate as the requirement, the Admiralty still refused to pay off the second half of the price. On the other hand the elderly Harrison, in his 70s by that time, still refused to give up.

Finally, in 1772 word of the feud even got to the new king, George III, whose predecessor originated the challenge. He himself evaluated the clock and was impressed with the results. Despite the king's favorable opinion, the Admiralty still did not pay. It took an act of Parliament in 1773 to force the second payment to Harrison.

Harrison was truly vindicated by Captain Cook, who took one of Harrison's clocks with him on his famed three years long trip. Upon returning he announced that the clock never erred more than 8 seconds during the whole voyage and declared it to be their "faithful guide". The challenge was answered, the ability to measure longitude by a clock was achieved. Harrison died in his 80s, a year after Cook's return, likely a satisfied and proud person.

It is fitting that the final chapter of the longitude challenge, that was instigated by the early years of sea travel toward the New World, starts with a contribution from the New World. Besides a tool to keep the reference time, there was also need for a common reference location. This was established in Washington D.C. in 1884 on the so-called Washington Meridian Conference, a gathering of the most important maritime nations.

The participants in the conference agreed to designate the meridian (longitude line) passing through Greenwich, now a suburb of London, to be the Prime Meridian, or the zero longitude line. The degrees of longitudes were to be measured to the east and west, up to 180 degrees. The 180 degree longitude (either east or west) became the international dateline. Using the dual possibility of measuring the longitude in degrees or in time, the prime meridian at Greenwich is 0 hours zero minutes and the international dateline marks 12 hours and zero minutes.

The international dateline is sometimes also known as the Sunday/Monday line. The line is roughly in the middle of the Pacific Ocean and when it is Sunday 12:00 noon on its eastern side, on its western side it is Monday at the same hour. The line is occasionally moved a few degrees for small distances to accommodate some of the island nations straddling the line. After all, it would be rather confusing to have some villages with different days on two sides of a street.

The time measurement of the longitude led to the time zones. Every 15 degrees in longitude, as we have shown earlier, amounts to an hour change in local time. For example, the location of the writing of this book, Los Angeles is at 118.24 degrees west from Greenwich, or the time here is 8 hours and almost 5 minutes behind Greenwich. When the Greenwich schools open at 8 am, the Los Angeles school children are still asleep at midnight.

The cities within any 15 degrees of longitudinal range all adhere to the same time. Hence the time zones are in hourly increments, except again on the Pacific, where in some strange cases half an hour time zones exist. Finally, some nations, like the United States, adjust the time zones seasonally, leading to Daylight Saving Time. This change is now national, albeit a few states do not adhere to the rule. This results in some temporary and localized confusion, but we digress.

Measuring local time at the poles is somewhat tricky because all longitudinal lines coalesce into a point. It appears that while standing on the pole, we can align ourselves with any time zone we like. For reasons of consistency, sometimes the North Pole is assigned to meridian zero, or Greenwich, while the South Pole is assigned to the international date line.

The next generation in the class of portable watches were the wrist watches developed early in the 19th century. Swiss watchmakers are credited with creating the first commercially available wrist watches. Incidentally here the ladies were leading and wrist-watches first became popular with them. With the advancement of aviation, the advantage of hand-free time keeping forced men also to abandon their pocket wonders. By the beginning of the 20th century wrist-watches were commonplace.

Despite the spectacular craftsmanship exhibited by the most famous watchmakers, the mechanical components limited the accuracy of measurements and humankind soon turned to physics for help.

7

July: Physical time

The first time measuring devices exploiting a physical phenomenon were the quartz watches. These clocks were based on a fixed frequency oscillation of quartz crystals and became wide spread in the last quarter of the 20th century.

Quartz crystals exhibit a regular oscillation when subjected to an electric current. Hence these clocks were equipped with a small battery that operates them for long periods (months or even years) of time. Since the first batteries had too small power to turn the arms mechanically, these clocks had to be digital. The time was reported by a liquid crystal display (LCD) with very low energy consumption.

Computers, GPS devices and other telecommunication equipments now all operate with an internal digital clock. Due to the advance of battery technology, today's digital clocks may now be combined with the conventional analog display of the hour, minute and even second hands. They may even have multiple faces for different time zones.

The desire to have even more accurate time measuring devices led to the atomic clocks. Atomic clocks are based on the spin rate of electrons in various atoms and their error is less than a second per million years. Such devices exploit the phenomenon that the nucleus and the electrons spin around their own axes. This is similar to the motion of objects in our solar system, after all, the internal components of atoms behave like miniature solar systems.

The spin of the electron and the nucleus both create a magnetic field and when their spin is in the same direction, the fields reinforce each other resulting in a higher energy state. When the spins are opposite, the magnetic fields cancel each other and the measurable energy state is lower. Incidentally the flipping of the spin of the electrons is at a highly regular frequency that is stable for as long as hundreds of thousands of years. Therein lies the perfect clock, the technical difficulties of measuring the energy states notwithstanding.

The specific atom used for this is the Cesium-133. The International System of Measurement (SI) standard defines the second as 9,192,631,770 cycles of the Cesium atom's electron spin changes. The importance of the atomic clock lies in the fact that the calibration of time measuring devices may be executed independently of an actual physical clock. Somewhat similarly to the naval clock challenge in the last chapter, where the role of Harrison's clock was to physically maintain and carry the reference time.

It appears as we have completed a cycle. We started with measuring time via nature, specifically the celestial cycles. The smallest separately named time unit, the second was one tick of the 86,400 tick long mean solar day. Finally we returned to nature again, and the second, our smallest humanly used unit, became the multiple of another natural cycle: 9,192,631,770 clicks of the atomic clock.

Having established the universal second, as this is called, we have a basis for a universal time. Universal time is the mean time at the Greenwich meridian and sometimes called the Greenwich Mean Time, or GMT. Since 1925, GMT is measured from midnight (0 hour) to midnight. Prior to that it was measured from noon to noon, but that time of the day is always associated with local celestial events that are hard to omit. Midnight is a less observable time, more acceptable for a standard.

Time has a fundamental role in physics. In the Newtonian equations of motion, the distance traveled by an object is the product of its velocity and the time duration of the motion. The longer the time duration, the longer the distance is, assuming a constant speed of travel.

Traveling with 60 miles per hour for one hour will take us 60 miles down the road. We may have done this in the past and surely it will have the same result in the future, banning an accident on the way next time. It appears that physical time is very definitive. A simple representation of this relation may be in a Minkowski diagram, such as shown on Figure 7.1 charting time versus the distance traveled.

Hermann Minkowski was a Lithuanian born immigrant to Switzerland and an instructor at the Swiss

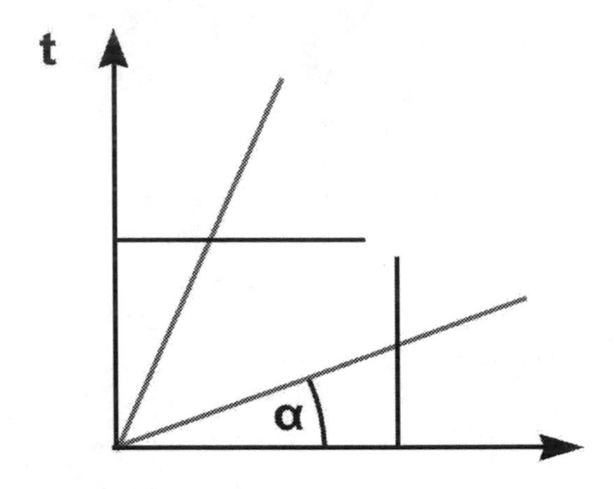

FIGURE 7.1 Minkowski diagram

Federal Institute of Technology in Zurich. He generalized the simple distance vs. time diagram of elementary physics into a time vs. space diagram. In such a diagram the three spatial dimensions are represented by the single horizontal axis on Figure 7.1. This simple looking generalization is very deep however; Minkowski coupled the space and time coordinates into the space-time continuum. While mostly Einstein gets credit for it, it was Minkowski's contribution that became instrumental in Einstein's work in relativity, the subject of our next chapter.

It is very simple to see that different speed lines result in different distances traveled in the same time interval. Incidentally, in this presentation, the lower the α slope of the speed line, the farther the object travels in a given time. For simplicity we assume a linear time vs. distance relationship, hence the lines are straight in the figure. That is of course not a necessity.

On the converse, the higher the α slope of the line the longer time it takes to reach a certain distance. This is somewhat counter-intuitive to our instilled sense, but it is true in a time vs. space coordinate system. Following the logic, an object whose motion is described by a line parallel to the horizontal axis on the figure, having a time-distance line with zero slope, has infinite speed, since it changes its spatial location in no time at all.

Finally, objects whose motion is on a line parallel with the time axis, shown by the vertical time-distance line, are just "biding their time" on the same spatial location. They are spatially stationary. Both of these extreme scenarios will be subject of further scrutiny later.

Newton's law works the same way independently of the direction of the motion. Let us consider using a negative speed quantity in Newton's law of motion example above, and the resulting distance will be also negative. The car would move backwards the same distance as forward.

If physics allows the object moving backwards, does it allow measuring time backwards? It appears it does with the help of carbon 14, a radioactive isotope of everyday carbon that is abundant in our atmosphere.

The so-called carbon dating technique measures the carbon 14 isotope content of any material containing carbon. An isotope is simply a variation of a certain atom with a different atomic arrangement. Specifically, the number of neutrons in the nucleus are different and the atom weight changes. On the other hand, the number of electrons remains the same, hence the chemical properties are unchanged.

The technique was invented by Willard Libby an American scientist only about 60 years ago and was rewarded by a Nobel price. He measured the ratio of the radioactive carbon 14 vs. the regular carbon 12 as about 1 per trillion. Furthermore, he also recognized that the radioactive carbon 14 will decay into carbon 12 by emitting neutrons with a half life of approximately 5730 years. This means that 5730 years after its creation, only half of the carbon 14 isotope remains, after 11460 years only a quarter and so on.

Since the chemical properties of isotopes are the same, carbon 14 couples with the atmospheric oxygen into carbon-dioxide, just like regular carbon 12 does. Plants will absorb atmospheric carbon-dioxide during the photosynthesis process and after the plants die, their carbon content does not change, except for the carbon 14 radioactive decay. When a fossil is unearthed, its carbon content can be measured and partitioned into carbon 12 and carbon 14. Depending on the ratio in comparison with the original 1 in a trillion and using the decay rate, one can establish the age of the fossil.

These are very small numbers; therefore the measurements are not exact. It is easy to lose a carbon 14 or two, among a trillion of carbon 12 atoms. The error of the method could be as much as plus or minus 30 years. Hence the accuracy of the dates so obtained may be only in centuries, but corroborating them with other historical and archeological data makes the method extremely valuable.

There is another limit of the technique, stemming from the fact that the decay rate of carbon 14 leaves only about 0.2 percent of it remaining in 50,000 years. That is only 2 in 100,000,000,000 atoms, simply too small to measure, limiting the dating by carbon 14 to less than 50,000 years. The measurement is delicate and requires high accuracy laboratory equipment, but objective.

Since we can measure time backwards to a certain distance via fossils, it is natural to raise the questions: Can we measure further back? What about measuring the age of non-fossil materials, such as rocks? The solution to these questions is similar to the carbon dating technique. We need to use the radioactive decay phenomenon, but with a material that has a much longer half-life and not carbon based.

Uranium, the most well known radioactive material and the medium helping the discovery of radioactive phenomena, comes to mind. The radioactive isotope uranium 238 decays into lead 236 with a half-life of 4.5 billion years. This is a very large time duration and the resolution of such dating technique would also be good, as it would lose only about 1 percent of its content in about 75 million years.

Luckily, uranium is widespread in certain minerals, most notably zircon. Zircon is also a perfect carrier of uranium because its melting point is so high that it survived long times during Earth's turbulent volcanic history. We found our clock to look back into time on an Earthly scale. One only needs to compare the lead content vs. uranium content of mineral zircon deposits to find its age.

With such uranium-lead measurements, Earth's age was determined to be about four and a half billion years. This was replicated by measurements of many rock deposits at various locations. Even meteorite rocks, occasionally found in Antarctica enclosed in pure ice layers, corroborate the age of Earth. This seems to imply that time is absolute, or is it?

8

August: Relative time

Albert Einstein, the famed German physicist and one-time student of Minkowski at the Institute of Technology in Zurich, revolutionized physics in the early 1900s. He essentially turned the time concept upside down, when he proposed that two persons observing the same phenomenon but moving at different speeds will perceive a different elapsed time, hence time is relative.

Einstein once illustrated time's relative nature by saying: "Put your hand on a hot stove for a minute and it will feel like an hour. Sit with a pretty girl for an hour and it will seem like a minute". The way to quantify relativity was presented in his theory of special relativity.

Einstein was famous for his so-called mental experiments in which he described hypothetical but plausible scenarios with physical components and contemplated the feasible outcomes. We will follow his footsteps by imagining a person sitting on a bench at a train station and a train standing directly in front of the person. To maintain a semblance of practicality, we will assume that it is a flat bed freight train car. We imagine that a young boy is bouncing a ball down the floor and back to his hand while standing on the car.

Figure 8.1 depicts the path of the ball.

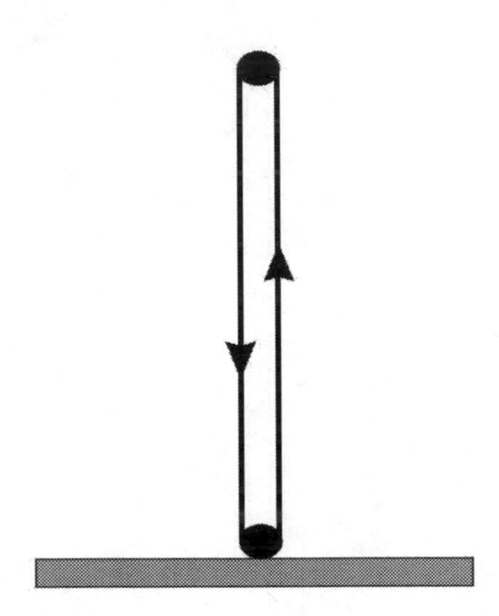

FIGURE 8.1 Time duration

Assuming that the height of the boy's hand when the ball is dropped is h, the ball is going to cover the path twice, once on the way down and back. According to Newton's law of motion, the distance traveled by the ball is

$$2h = s \cdot t_p.$$

Here the time of completing a bounce measured by the

player is t_p and s is the speed of the bouncing ball. It is assumed to be the same going down and bouncing up, which is obviously a slight approximation of the real physics involved. In reality, when the ball is on its way down it is aided by gravity increasing its speed, and on the way up the same effect works against its speed. We will assume that the effect on the small distance scale is negligible and the constant speed applies for our mental experiment.

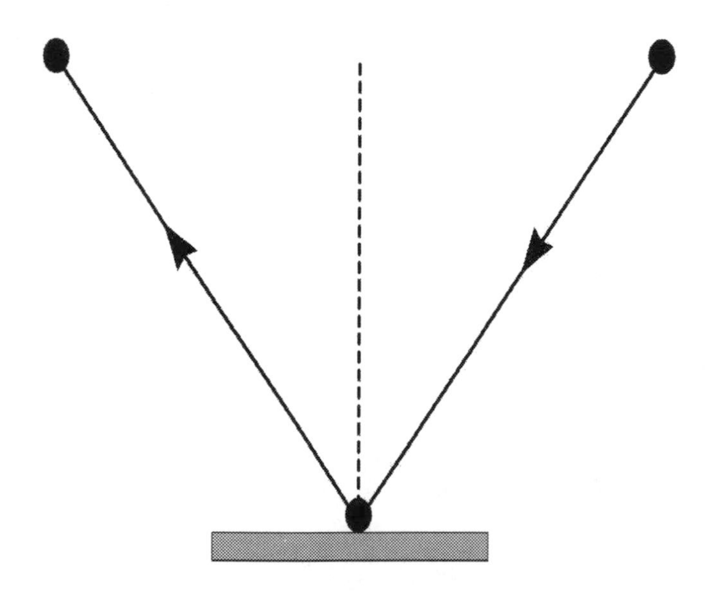

FIGURE 8.2 Time dilation

We now will stretch our imagination a bit and envision the train moving in front of the stationary observer on the bench. Figure 8.2 sketches the moving event. When the moving participant drops the ball it is somewhat to the right of the bench of the observer and by the time the ball hits the ground, the train car is directly in front of the observer. Finally, when the bouncing ball reaches the boy's hand again, the train has passed to the left.

The ball has traveled on a path along the two distinct slant lines with length d and according to Newton's law of motion the relation is

$$2d = s \cdot t_o.$$

Here t_o is the time of the bounce according to the observer on the bench.

Let us now compare the two simple equations. Since the distance d of the slant line is visibly longer than the vertical distance h, the left hand side of the second equation is bigger than that of the first. Hence the right hand side of the second equation must also be bigger.

The speed of the ball (s) does not change between the two experiments, after all, it is the same experiment. Therefore the only explanation is that the time of the observer is longer:

$$t_o > t_p.$$

We arrived at a very interesting phenomenon. For the boy bouncing the ball on the train there is absolutely no difference between the two events. If the train is moving on a very smooth rail, like the high speed trains of Japan or Europe, he may not even realize that the train is moving. He is part of the moving system; he does not detect any difference.

For the external, stationary observer, however, the train is obviously moving. The simple equation above implies that the duration of the event according to the stationary observer is longer. In essence, his clock is ticking faster than the clock on the train.

Considering the fact that they are observing the same event synchronously, this scenario is somewhat counter-intuitive, it would feel more sensible the other way around. For the fast moving participant of time, one would expect the time speeding up, not slowing down. The synchronicity in our experiment was assured by the common, constant velocity of the experiment and the same, single instance of a bounce of the ball.

This paradoxical phenomenon was called time dilation by Einstein. The magnitude of this discrepancy depends on the velocity of the moving train. The higher the velocity, the bigger the distance d compared to the h and the larger the time discrepancy. To quantify the amount of time dilation, Einstein created the intriguing formula:

$$t_o = \frac{1}{\sqrt{1 - \frac{v^2}{c^2}}} t_p.$$

The c is the speed of light and the v is the speed of the traveler, or the train in our case, not the speed of the ball: s.

This formula very nicely quantifies the effect. If the traveler's speed is negligible compared to the speed of light, the relativity term is almost one, consequently

$$t_o \approx t_p.$$

There is no quantifiable time dilation.

Now let's see if we could put it into the perspective of everyday circumstances. Consider a speed of 360 kilometers per hour, or 0.1 kilometers per second, a plausible speed for a bullet train. With the speed of light at 300,000 kilometers per second, the relativity term becomes about one tenth of one millionth above one.

An event transpiring on a speeding bullet train lasting one second in duration for the traveler on the train would appear 1.0000001 second long for the stationary observer on the platform. That time dilation is certainly not easily, if at all, measurable even with the atomic clock resolution of the last chapter.

The relativity term in Einstein's formula implies that the traveling speed at which this phenomenon would become measurable must be very high, close to the speed of light. We are very far away from building spacecraft capable of speeds close to the speed of light, the closest we get is the space shuttle.

The space shuttle is traveling almost 300 kilometers above the Earth and with 28,000 kilometers per hour velocity. To put it in perspective, about tenfold increase of that velocity would put it into the neighborhood of 300,000 kilometers per hour. But the speed of light is that many kilometers per second, so another almost 4,000 fold improvement in speed would be necessary. The space shuttle, the pinnacle of human velocity, still travels about 40,000 slower than the speed of light.

The time dilation endured by an astronaut during a daylong journey on the space shuttle would be about 0.000025 second, that is somewhat difficult to measure. But, surely we can create an atomic clock to keep the Earth time reliably. Hence the more difficult question is whether we can create a clock that would actually detect the time dilation.

Could we detect the effect of time dilation on the space shuttle astronaut? The effect of the 0.000025 second time dilation is likely undetectable. How about sending an astronaut into an Earth orbit with the space shuttle for approximately 55 years? The time dilation would be somewhere around a half second, a time amount that is measurable by everyday clocks.

Demonstrating the effect of half a second time dilation on the life of the astronaut during a 55 Earth years long shuttle trip, however, still surpasses the means of humankind. It appears that despite the plausibility of the time dilation concept, we have difficulties to produce a detectable demonstration of it on the human scale.

A detectable demonstration of the time dilation phenomenon ultimately has been done by accelerating subatomic particles in scientific experiments in large scale particle accelerators. When unstable subatomic particles move with speeds approaching the speed of light, their detectable lifespan is much longer than it is supposed to be. The only explanation is that their time has slowed, as predicted by the time dilation concept.

Another aspect of times's relativity may be demonstrated by even the one day long shuttle trip. The space shuttle travels about two thirds of a million kilometers during an Earth day. During that time the astronauts orbiting Earth would see a sunrise in every 90 minutes, or 16 in each 24 hours. We of course see only one in every 24 hours on the Earth.

This demonstrates the phenomenon that the duration of an event, in this case the time between two sunrises, depends on the observer's frame of reference, opening the door to a very interesting dilemma.

9

September: Twin dilemma

The fact we just established, that for a traveler with high velocity time will slow down albeit the traveler may not notice it, leads to the famous twin dilemma.

That is of course again a mental experiment, since we do not have the means to make a person travel with high enough velocity making time dilation really measurable. Let us, for the sake of the following discussion, assume that we can. Consider one twin being sent on a trip aboard a spacecraft with a velocity near the speed of light. Let us imagine that the twins are a girl and a boy, and the girl is traveling, for the sake of easy narrative.

Just like the person staying on the station platform observing the fast train moving by, the twin boy is staying Earth bound. In our mental experiment the twin girl leaves Earth on January 1st, 2011 and from the boy's earthly life 10 years have gone by, before the spacecraft returns on 2021 January 1st.

Let us imagine that the girl travels with three fifths of the speed of light. The relativity term in Einstein's formula of the last chapter becomes 5/4 and the time dilation becomes 25%:

$$t_o = \frac{5}{4} t_p = 1.25 t_p.$$

Since

$$10 = 1.25 \cdot 8,$$

the traveling twin's time duration will only be 8 years. This is rather straightforward, once the time dilation concept is accepted. The question is, how to interpret this in connection with space. We can use Minkowski's diagram to view the paths of the twins on Figure 9.1.

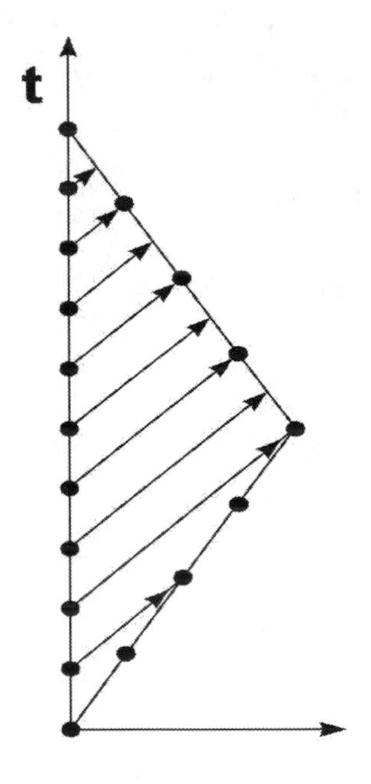

FIGURE 9.1 The twin dilemma

On the figure the horizontal axis is again the space direction. The motion in this direction will be measured in light years. Note, this is a distance, not a time unit. The vertical axis is the time and that will be measured in years. Each pair of bullets on the figure represent a year, 8 years for the traveling and 10 for the stationary twin. This resembles the different number of bullets representing the Earth vs. Venus years on Figure 1.2.

The home bound twin never leaves the vertical axis of the space time system, in other words he only travels in time. On the other hand, the traveling twin has an outward path and a return path in space. The diagram shows, that the traveling twin has returned to the home bound twin to complete the journey.

The outward spatial distance, when one travels at speeds approaching that of light, may be measured in light years, as is customary in astronomy. We divide the traveling twin's timeline into 8 equal segments according to her clock, the return point is half-way out, 4 years away. Since she is traveling with 3/5th of the speed of light, her spatial distance is $12/5 = 2.4$ light years, represented by the tip of the triangle.

On her return trip, she covers the same distance, hence she travels 4.8 light years in distance overall. Upon return, she will find her brother aged 10 years, while she aged only 8, a flattering ratio between men and women (and a reason why we had the girl travel). Joke aside, her vital characteristics would be of a person 2 years younger than her brother, if we had an absolute way of measuring them, which we do not.

How can we synchronize the time between the twins? We return to Minkowski's diagram and recognize the fact that a 45 degree line would represent an object traveling with the speed of light. That is, in moving one unit to the right on the space axis (one light year) it moves one unit up on the time axis or a single time year.

On Figure 9.1 the slant line with a bigger slope than 45 degree represents the traveling twin's outgoing timeline. Following the logic of the last chapter about time-space coordinate system, the higher slope means it takes longer than one time year to travel one light year distance, adhering to the fact that the spacecraft is traveling with less than the speed of light. The return trip, is represented by a backward sloping line. Note, this does not imply back in time, only back in space.

Let us now draw a 45 degree line from every year of the stationary twin toward the timeline of the traveling twin, as shown on Figure 9.1. The 45 degree lines may be depicting messages sent via electromagnetic waves (that do travel with the speed of light) by the home bound twin to the traveling one. On the figure the timelines of the two twins are discretized according to their respective clocks. Hence the traveling twin's timeline consists of less bullets than the stationary one.

On the outgoing path, the traveling twin is speeding away and receives less messages than were sent. This is less intuitive, even in light of the time dilation. The effect is because the outgoing spacecraft is moving away in the four dimensional space time. Some of the first messages do not catch up with the traveling twin until the turnaround point.

On the return path, however, the number of messages is increasing and more than one yearly message is received by the traveling twin in her year. The reason for this is that some of the messages that did not catch up during the outgoing path are now received. The returning ship, moving toward the source of the messages is intercepting more than one message per year.

All the messages sent by the home bound twin are received by the traveling one, but their frequency has changed. The temporally equidistant yearly messages arrive in unequal temporal distances at the high speed aircraft. One message is received in every second year going out and two per year on the return route.

This mental exercise can be carried further. Let us assume that the traveling twin is leaving on a spacecraft with the speed of light. On Minkowski's diagram this would mean a 45 degree line. Hence, on the outgoing trip none of the messages would catch up with her. On the return trip, represented by a negative 45 degree line, all messages would be received in regular time intervals, albeit shorter than the sending person's time intervals.

Let us now turn the role of the message sender around and assume that the traveling twin is sending yearly messages back to Earth. Even though she will send them at regular yearly time intervals according to her clock, the Earthly twin will receive them at different times. Since the traveling twin has only eight new years on her clock, there will be only eight messages received on Earth.

Note, that the twins' dilemma is equivalent to the mental exercise of the train of the last chapter. Here the event is the duration of the time from the moment the spacecraft left Earth, until its return. The participating twin is the traveling person and the home bound twin is the observer. Their respective time difference is the manifestation of the time dilation.

A very interesting aspect of this dilemma is the simultaneity of events between the two twins. Obviously the moment of departure is simultaneous from both of their perspectives. The stationary twin watches his sister's face in the window of the spacecraft and she looks at him standing on the ground. The arrival is a simultaneous event of their embracing each other.

What happens in between is, however, more difficult to judge. Let us mentally draw horizontal lines on Figure 9.1. The lines would connect two points in the life of the twins. They are the lines of simultaneity between two distinct space-time locations, although the stationary twin only moves in time, not in space. The events happening in their respective lives at the two points are simultaneous.

Let us consider each point on the timeline of the twins as an event in their lives. It appears that the traveling twin would have more events in her life, since the slant lines of her timeline are obviously longer than the straight line of the stationary twin. This is in conflict with the time dilation principle proposing that if anything, the stationary twin should have more events in his life. After all, his time duration is longer.

This conundrum lies in the fact that both lines contain infinite number of points. The famous German mathematician, Cantor proved that a one to one correspondence could be made between any two line segments of different lengths; their order of infinity is the same. Hence there are the same numbers of infinitely small time instances on both lines, although certain events of finite time duration may be different in length.

We can see this if we view the phenomenon in a physiological, as opposed to physical sense. We know that time is an intrinsic component of our physiological life, the cells and organs of our body react to the passing of time. The metabolic rate and all biological processes in the traveling twin's body would slow down and she would age less.

The metabolic rate has a deep connection with the lifetime of all living creatures, including humans. It is a known fact of biology that the smaller an animal, the faster its metabolic rate. One can say that smaller animals live faster. It is almost as if their internal clock was ticking faster.

On the other hand, the larger the animal, the slower its metabolic rate is. This fact is manifested by the longevity of elephants, whose life time reaches into seven decades, and sometimes even more. It appears that the internal clock of large mammals is running slower, as if time dilation had occurred.

Where do humans belong in this range? The average human age is certainly comparable to that of the elephants, well into the 70s. What about our internal clock? We considered the heart beat as a reasonable small unit of time earlier, so let's quantify our metabolism with it. A little bit of arithmetic with a calculator yields the fact that a human heart of a person who lived to the age of 75 beats almost 3 billion times, assuming a heart rate of 72 beats per minute.

This number is quite large. Our comparison mammal, the elephant, has a total number of heartbeats of only 1.2 billion, using the average heartbeat of 30 per minute measured by biologists. It is clear that the elephant's body is many times that of any human and still lives about the same length of time. It appears that the elephant's lower heartbeat rate is a useful thing, hence time travel would also be such. Or would it?

A consequence of the slower time is that the traveling twin lived less. Since her life span aboard the spaceship covered only eight years, the amount of learning, pleasure and experience gained during her trip is only eight years worth. Yes, she arrives younger, but less experienced or mature, and possibly less satisfied. Combine that with the likely confined nature of a space capsule traveling at such high speeds (assuming we can build one) and the question arises: was it worth it?

The Earth bound twin has experienced all the joys and pleasures of 10 years of life fully, yes with commensurate aging, but probably without remorse. On reflection, boys might be more of this kind after all, less concerned about aging and more interested in living life to its fullest extent. We are on a shaky philosophical ground and becoming stereotypical, so we shall leave it at that.

If we carry the experiment further and assume a traveling time of, for example, 80 years for the traveling twin, another interesting situation appears. The Earth bound twin may be long gone, after all, 100 years passed by since their separation at the start of the trip. Even if the trip started in their toddler years, the traveling twin, by now herself an elderly lady, comes back to Earth in the future.

Yes, she has traveled in time, specifically into the future because had she stayed, she would not be alive. This, now rather philosophical concept is worthy of some discussion. How far can one carry this? In our example, we still need the traveling twin to be alive to arrive in the future, so we are somewhat limited. However, the closer one gets to the speed of light the bigger the time dilation becomes, as we have seen it in the last chapter. Hence, by traveling with 95% of the speed of light, one can travel thousands of years into the future.

Then again, taking the metabolic rate slowdown a bit farther, we can imagine the traveling person to be put into a hibernation state, much like the bears in their winter caves, and the future is almost infinite. This is now the realm of science fiction films, many showing exotic looking chambers from which astronauts emerge upon arriving at a faraway planet.

That brings the idea of actually decoupling time travel from the spaceship and the speed of light, by suspending the life of the living organism traveling in time. We arrived at the realm of cryogenics, maintaining a body in a reduced body temperature hibernation state until a future time. This could be another facet of the twins' dilemma, one frozen at the time of birth and thawed when the living twin is near death, but which one.

If we had the understanding of how the brain works, where our elusive soul, personality and life experience are stored, the hibernated twin's body could be loaded up with the life experience of the living one and the time of their combined life extended.

This is now one step away from cloning, an intriguing if controversial topic that we do not go into. One thing is sure, we humans would go to the extremes in our desire to cheat, prolong, or even stop time.

10

October: Stopping time

Einstein's formula introduced in Chapter 8 also implies that when the traveling velocity reaches the speed of light, time stops. This seems like an incredulous scenario. Why would this happen at the speed of light? One could possibly accept that to happen when the traveling speed is infinite.

This certainly would be more palatable for our everyday thinking, after all they both, infinite speed and infinitely small time progression, are just theoretical concepts. We do not anticipate reaching either one ever. We learned early on in our mathematics studies, infinity is something we never reach.

According to the formula, however, the clock would completely stop when the traveling speed would reach the speed of light. Time would be at a standstill. The inevitable conclusion is that the speed of light is unreachable. If we can never reach that speed, then time will never stop either, so we regained the sanity of our intuition and breathe with a sigh of relief.

Einstein proved just that. The speed of light is the ultimate speed, even though it is finite. This fact was the subject of centuries of debates and scientific arguments. One of the first methods to quantify the speed of light was conceived by Galileo Galilei in the later years of his life. He devised an experiment where two persons carrying lamps equipped with shutters were placed very far, several kilometers apart. One person would signal a flash of light by quickly opening and closing the shutter. The second person, upon seeing the light, would act similarly and send a flash back.

Galileo did not live to see his experiment that was finally conducted about 25 years after his death, commissioned by the Florence Academy. In the actual experiment the two participants were executing a flash exchange, then moved farther apart and did so again. Several of these were executed at increasing distances, and to the consternation of the participants the time elapsed between the flashes did not change, no matter how far the lamps were from each other.

This seemed to imply that the speed of light was infinite, or at least, much faster than the experiment could detect. From the actual distance, they computed a lower limit in the neighborhood of 10,000 kilometers per hour, a rather formidable speed. Considering also the reaction time delay of the participants in opening the shutter after seeing the light, they concluded that it was likely even higher.

That estimate was refined not much later by the Danish astronomer Ole Roemer, once an assistant to the famous Tycho Brahe, whose observations led Kepler to discover his laws. Notice the rich collection of historical characters in a small time span of the second half of the 17th century.

Roemer spent several years observing the motion of Jupiter's moons, and their relative appearance from Earth. He understood the geometry at hand, that they are circling Jupiter, which was not in question by his time. The universe rotating around Earth was, however, still a widely held belief and the famous characters of the last paragraph were major contributors in dispersing that myth. Roemer became an expert on one of the moons of Jupiter, Io, so much so that he was able to predict one of its eclipses to the minute.

More importantly, his observations resulted in his estimation of the speed of light as approximately 190,000 kilometers per second. A huge jump from the Florence estimate just a few years back, and a respectable number in view of the now known actual value of about 300,000 kilometers per second. Even the dilemma of finite vs. infinite was put to rest with that estimate measured from repeatable celestial phenomena.

That, however, does not solve our problem of the speed of light's boundary nature, implied by Einstein's equation. So, how did Einstein arrive at that? As mentioned before, Einstein liked to immerse himself into thought experiments. In his famous mirror experiment he imagined someone traveling in a spacecraft with the speed of light and holding a mirror in front of his or her face. In stationary circumstances we all know that we can see the reflection of our face in the mirror. Einstein was wondering whether we would still see it when traveling at the speed of light.

After all, if the light leaving the face of the traveler was proceeding with the same speed as the traveler, i.e. with the speed of light, it would never reach the mirror. Hence the traveler at light speed would not see a face in the mirror. This did not fit with the mechanical relativity principle of Galileo which was widely accepted by that time.

Galileo has posited a mechanical principle of relativity centuries before Einstein, stating that someone traveling at a constant speed should not be able to decide whether the motion is forward, backward, or even stationary. Galileo used an argument with respect to a hypothetical ship that we will translate into our high speed train of the earlier chapter.

Let us imagine someone sitting in a cabin of the train with windows shuttered, and tossing a book to a person sitting across. If the train's motion is extremely smooth, the person has no way of knowing whether it moves or not, and if it does, which way. Furthermore, there is no difference in the effort required to toss the book independently of the direction of motion.

This was surely contradicted by Einstein's mirror experiment. Something was not right. Einstein pondered the problem for several years until he arrived at the question that resolved the problem. He asked: the speed of light is 300,000 kilometers per second (by that time the value was known), but relative to what? The answer came to him that it is relative to the observer.

This meant that independently of the origin of the light, it always travels with this speed relative to the observer. The traveler with the speed of light will see a face in the mirror, because the light leaving his or her face travels with the speed of light relative to the person. It will safely reach the mirror, bounce back and return with again the speed of light relative to the traveler.

The constant nature of the speed of light relative to any observer, independent of the direction of motion is a hard to swallow idea. It seems like it is easy to contradict it with everyday examples. After all, it is well known that when two cars collide head on by each going 60 miles per hour, the resulting damage is proportional to one car crashing against a brick wall with 120 miles per hour. So aren't relative speeds of moving objects are added to each other?

As it turned out, that did not apply in the case of the light. It was not a moving object, it was light. But, how could its speed be a limit to velocities of all moving objects? Why can't anything go faster?

The answer's details may lie a bit beyond our focus, but we'll attempt to formulate it simply. The issue boils down to the equivalence of mass and energy. The energy required to move an object of resting mass of m with a velocity of v is

$$E = \frac{1}{\sqrt{1 - \frac{v^2}{c^2}}} mc^2.$$

Here the now familiar relativity term is 1 when the velocity is zero, hence in the case of a stationary mass, we arrive at Einstein's most famous equation in its well known form of

$$E = mc^2.$$

The more important case for our discussion is when the velocity equals the speed of light $v = c$. In this case the relativity term becomes infinity. Consequently, the energy required to move an object with a finite mass at the speed of light is infinite, a physical impossibility. Hence Einstein's proposition that a moving object cannot reach the speed of light is correct.

For historical fidelity it is proper to mention that Einstein was not the first to write this equation. While Einstein's original paper in 1905 contained the relativity term we used in the prior parts, the energy-mass-speed of light trichotomy was first written by Max Planck, the founder of quantum theory, in 1907 in a slightly different form.

Einstein himself only wrote the simplified formula almost 4 decades later, in his years of living in the United States, in an article he wrote for the magazine Science Illustrated. That was the time when his ideas and theories became famous and the formula itself a household phrase worldwide.

Time's relative nature leading to the twin dilemma and the speed of light's limiting nature seem to conflict each other. Einstein himself was not happy with the name of theory of relativity in this regard. He contemplated to name it the theory of invariance instead, when dealing with the speed of light. The two are, as we saw it in the time dilation formula, however, inseparably intertwined.

Einstein was concerned about the relativity phrase catching fire in popular imagination and getting misused. Sure enough, soon "everything is relative" became an everyday expression and used inappropriately in social and economical contexts. Einstein's fear was justified, but the avalanche could not be stopped and the use of the phrase proliferated.

In truth, we must clarify that the speed of light is not the ultimate limit of speed for everything. There are phenomena that move faster than the speed of light, but note the careful wording. One example often quoted is moving a laser light beam across a surface. The spot of the laser light on the surface is moving faster than the speed of light, because its original speed of light progression is compounded by the lateral speed of the motion.

But the crux of the matter is that the spot does not have any mass. Mass is the critical component of this discussion. An object with no mass requires no energy, hence can reach or even exceed the speed of light.

An actual physical demonstration was also executed by scientists in the last decade. A chamber filled with cesium vapor, the extremely accurate time measuring material we met earlier, was bombarded by laser impulses. The speed of the laser impulse passing through the chamber exceeded the speed of light - a spectacular breakthrough over the ultimate speed.

According to the observations, the laser beam left the chamber before it even finished entering. This sort of even implies reversing time. Having stopped and now reversed time, it is natural to ask, can we travel back in time?

The idea of traveling backward in time is particularly alluring. Such a possibility, however, leads to peculiar circumstances, called temporal paradoxes.

11

November: Grandfather paradox

A temporal paradox is a scenario in which the time traveler can cause events in the past that would prevent the possibility of the time travel to occur in the first place. The most well known is the paradox in the title of this chapter, in which someone travels back in time and kills his (let us now have a boy traveler) grandfather before his father was even conceived, let alone born. The paradox lies in the fact that in that case the father of the traveling person could not have been born and the person himself neither. No person, no time travel.

There are several philosophical arguments to counter this paradox and allow the possibility of backward time travel. One of them is called eternalism, that utilizes the space-time concept of the past chapter. It states that since time is just another dimension in our four dimensional world, all future and past time coordinates (i.e. events) are also there. Just like all the spatial locations are already there before we traveled there, so are temporal locations of the past and the future.

An opposing philosophical school is that of the presentists, who posit that only the present exists in the time dimension. In their view the past is gone, the future is not here yet, hence there is no other way to travel in time, but waiting for tomorrow. While this seems more natural to our thinking, contemplating the possibility of time travel is so intriguing that many movies were made about it.

A notable and very successful one was the 1985 film titled "Back to the future". In this movie the time traveling high school student goes back to his parent's high school days to correct a life-long personal conflict between his father and one of his high school mates. The comedic effects of the mother taking a strong, but not maternal, liking to the son and other similar confusions made the movie worthy of two sequels.

Another comical aspect of time travel is arriving in a time machine at a spatial location that is not the nice laboratory of the scientist who built the machine. Arriving, for example, in the middle of a medieval battle, led to many other hilarious film conflicts.

Time travel also has a long literature history. Perhaps the most famous, but by far not the first of them is the H. G. Wells story titled "The time machine" that was published more than a hundred years ago. Wells' phrase, the time machine, has since become commonplace.

Let us look beyond the grandfather paradox and view it in the more generic sense of influencing events that have a bearing on the future, i.e. changing history. Unless the presence of the traveler from the future is absolutely unnoticeable, it could result in unforeseeable consequences. This is the so-called butterfly principle at work that claims that every wing wave of the butterfly has ultimately magnified consequences and result in changes to the future.

Future's extreme sensitivity to the past leads to another logical argument that is very plausible. Because of the possibility of modifying the past by traveling back would make the present already pre-defined, we appear to be living in a closed time loop. Whatever happened has already happened.

This train of thought does not explicitly preclude the possibility of traveling back in time. It just states that even if someone travels back in time, he has already traveled back in time, i.e. part of an already happened past.

The closed time loop concept is also in the popular culture. The frequently occurring deja vu scenario, one reliving an event that has already happened earlier, is well known to most people. An interesting exploitation of the idea was in the movie titled "Ground hog day" in which a TV reporter had to relive the morning of the ground hog's appearance from its hibernation again and again. In this case, the reporter had some behavioral deficiencies that when rectified, enabled him to get out of the loop and continue his life.

Deeper thinking about the grandfather paradox can lead to a logical as opposed to temporal paradox. In this form it is called causality paradox, whose archetype example is the well known chicken and egg paradox posed in the question: Which came first, the chicken or the egg? There is a similar cause and consequence dilemma here, like in the grandfather paradox.

Aristotle, arguably the most famous classical philosopher, addressed the problem already in the 4th century BC. After much debate, he decided that both an original chicken and egg must have coexisted. This, on its face wise answer actually did not resolve the problem. There are of course two other solutions, deciding in favor of either side.

The creationist approach to the problem is based on the Bible describing God creating the birds (we assume the ancient chicken is a descendant of birds) and directing them to multiply. There is no mention of the means of multiplication and of any eggs. Hence the decision based on the Bible comes down on the side of the chicken.

The evolutionary philosophy also favors the chicken. It assumes that the chicken evolved from birds that may not have laid eggs, but during their own evolution developed the technique of laying eggs. Overall, it seems like the chickens are winning. Humor aside, the argument itself is not resolved.

At the core of any causality paradox is a circular reference. The existence of one assumes the existence of the other, and vice versa. Circular references often lead to contradiction. Another classical philosopher named Epemenides, who lived on the island of Crete in the second century BC, created a timeless example of a circular reference. He supposedly stated: "All Cretans are liars".

The paradox is rooted in the fact that Epemenides

was also a Cretan. Hence, according to the statement he was also a liar. But then his statement would not be true and Cretans, including him, would not be liars. But then his statement would be true and he, after all, would also be a liar... And on and on...

Variations of this were created by modern philosophers, such as Bertrand Russell, whose barber paradox goes like this: there is a town with one barber and the rule of law is that he shaves all the men in the town who do not shave themselves. The question arises: who shaves the barber?

Well, if the barber does not shave himself, he must obey the town rule and get shaven by the barber. But if he shaves himself, then according to the rule he cannot be shaved by the barber. It appears that the barber cannot shave. The paradox is of course in the fact that the barber is both a citizen of the town as well as the lone barber. As all paradoxes, they have something ambiguous or even contradictory in their propositions.

We strayed away from the actual temporal aspects of the grandfather paradox, but these logical paradoxes seem to show a way for arguing the case. Let us first consider the following train of thought. All the events in the past had already happened. If anyone ever visited the past, from our present, or from our future, it should have been part of our historical record of the past. Since there are no such records, it appears that it will never happen in the future.

There seems to be a loophole in this argument in the fact that its reference frame is today's present. That should not prevent someone from the year 2200 traveling back to the year 2100, since both of those dates are in our future. But the relative relation between the grandson and the grandfather makes the actual date irrelevant in this argument. The grandfather's siring the son, the grandson's father is in the already happened past, so the same argument will apply to the grandson of the year 2200 as for the grandson of today. This seems to decide the grandfather's paradox on the side of impossibility.

Time travel into the past is prevented by some technical arguments as well. There is the reasonable sounding argument that one cannot go back in time before a time machine was ever built. Furthermore, it is a safe assumption that any time machine will need a start and destination equipment. Since we have not found any destination time machines in our time yet, time machine probably has not been yet invented in the future either. Hence it appears strongly that one cannot travel back in time and exercise the grandfather paradox.

There is a possible way out of the grandfather paradox. The so-called multiverse theory proposes that we are living in one thread of reality, but there are multiple threads with a common connection in the past. According to this theory, when the boy would travel back in time and kill his grandfather, the thread accommodating that event would be different than the one where the boy will be born. This would be something like time having multiple dimensions, coexisting in the physical space.

This co-existence of multiple time dimensions leads to the mysterious phenomenon of the time slip. Some people alleged to have "slipped" back in time while being in the same physical space. Among the incredible stories are two English women back in the 20th century who claimed to have slipped back in time while walking in the gardens at Versailles, and met with characters of the French revolution 200 years earlier.

Then there was the case of two British couples traveling through France on holiday, who allegedly encountered a lovely old-time hotel on their way south. They stayed overnight, took some pictures and made a mental note to stay there on the return trip. Curiously, however, they could not find the place on the road and upon returning home their photos taken of the old hotel were also missing. Their story actually made its way onto the British television.

There must be something with British folk traveling in France that they are so prone to time slip. It appears that our quest to resolve the Grandfather's paradox led us into the paranormal territory of the mind, so it is time to abandon the topic.

But there are scientific arguments as well. Stephen Hawking, probably the most known scientist of our times, conjectured that the physical laws are such that they prevent time travel on all, but subatomic scales. This seems to even apply in both directions, we saw earlier that the theoretically possible time travel into the future is practically impossible on a human scale, due to the forbidding energy requirements.

Another argument against traveling back in time is the presence of temporal barrier events in our past history that we cannot overcome. Earth's early volcanic history and the creation of the continents is certainly one such barrier. We cannot travel back to a place in time before it was created.

The most spectacular temporal barrier is the Big Bang, that certainly appears to be an impossible barrier to travel back beyond. This is despite the consensus that there was time before the violent physical event of the Big Bang. The theory now called the Big Bounce proposes a cyclic model of an oscillatory universe, where an expansion phase of the universe is followed by a deflating phase.

In this theory the Big Bang was not the first cosmological event, it was actually the result of the collapse of the previous universe. If this theory is true, time has eternal cycles, leading to the December of our cycle of contemplating of time.

12

December: Eternal cycles

We have seen some truly grandiose cycles of time in the preceding chapters. It is a natural desire of the human mind extending them further, describing infinitely long cycles of time leading to eternity.

Pondering time and especially eternity was a favorite pastime of philosophers since ancient times. Aristotle, famed teacher of Alexander the Great, and contributor to the paradoxical issues of backward time travel, had a lasting influence on this topic.

Aristotle argued that time is eternal. His argument was based on the assumption that time is a series of moments. The present moment is the beginning of the future and the end of the past, hence each moment has a predecessor and a successor moment. If there was a very first moment without a past, the beginning of time, that would violate the above assumption. On the other hand, if there was a very last moment, the end of time, this would also contradict the original assumption. Hence time must be infinite in both directions.

Infinity as a concept emerged in antiquity, in philosophical, algebraic and various geometric forms. A geometrical loop, such as a circle, is a way to describe infinity. One can repeat it over and over again.

FIGURE 12.1 The Ouroboros

Infinity and eternity were depicted by the ancient symbol of the Ouroboros shown on Figure 12.1 and already seen in some Egyptian artifacts as early as 1600 BC. The name itself was given by the Greeks; the word means tail-eater. By now the word means all the various symbols that are conceptually similar. The serpent, or snake eating its own tail symbolized the continuity of life, its cyclical nature of birth and death.

The symbol transcended cultures and religions. It also appeared in the Roman mythology where it represented the God of time, cycling through the months of the year and repeating the cycle all over again, i.e. creating a loop. There are events that are cyclical and easy to visualize in terms of the Ouroboros. The motion of planetary systems around us, the yearly seasons, the days and nights, the high and low tides, are all of this class. From our perspective they are infinitely periodic and will never change.

But there are cyclical events that are inherently decaying or amplifying in time. The deterioration of the cellular health of our body during the years and our knowledge gained during our life span are the most personal examples of this category. Let us find a geometric manifestation of infinity involving such cycles that will describe the road to temporal infinity, eternity.

A geometrical object manifesting such change is the spiral. Its mathematical definition is simple, it defines a point traveling away from the origin by a certain linear and angular speed. This means that the distance of the point from the origin and the distance on the arc of the curve are increasing simultaneously. It is easy to imagine that the spiral curve represents time and its geometric infinity represents temporal infinity.

Our time spiral, shown on Figure 12.2, is infinite in two directions. This is a perfect geometric shape aiding our visualization of time's evolving cycles and its infiniteness, both in the past and in the future. It goes

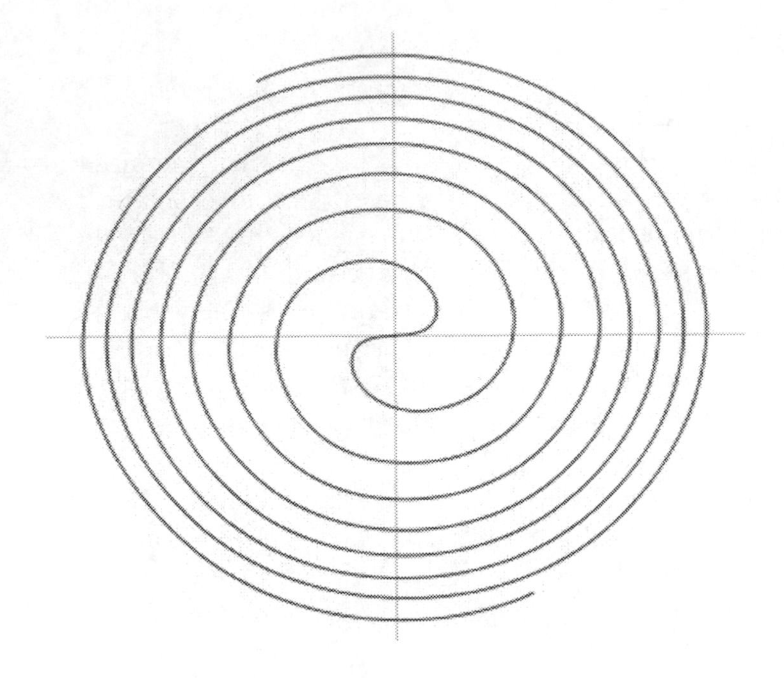

FIGURE 12.2 Time spiral

through the origin of our coordinate system and one can perceive that point as being the present. The axis of the time dimension of our four dimensional space-time coordinate system is this spiral curve.

We can assign the calendar years to each repeated 360 degree turn of the time spiral. The other 5 days and some hours to complete our proper year will be overlooked for the sake of the analogy. Measuring the loops in degrees, we represent the constant length of each calendar year cycle. The ever increasing length of the spiral loops in both directions may represent certain time related phenomena.

Let us assume that the time spiral represents our personal life, its turns our years and the origin the present. All events of the future lay on the forward side of the spiral and all moments of past on the backward side. All the events in the past we remember (well most of them anyway), and the events in the future we may anticipate, but cannot really predict.

The increasing length of the turns in the future may represent the perceived increased pace of passing time as we get older. We are all familiar with the memories of the endless summers of our youth and the ever faster passing of the working years. This may be related to the phenomenon that a faster journey over a bigger distance appears to be longer than a slower journey over a lesser distance, despite the fact that the duration of both journeys were the same. Perhaps we were living faster when we were young.

This feeling is sometimes called psychological relativity, and is truly only in our minds. Our body has an internal neurological clock, the circadian rhythm that seems to know that we live in 24 hour long cycles.

On the converse, the increased length of the cycles of the past may represent our ever increasing amount of accumulated memories. This seems like a certainty, after all we do learn more and more things through the years. Whether we remember them all is another issue, we seem to have forgotten a lot that we learned in school.

Thinking in this spiral time world, the concept of time travel, subject of an earlier chapter, may be interpreted as jumping to another point on the spiral, time traveling without staying on the spiral curve. The curved nature of the spiral implies that there are longer and shorter "time" distances that one can travel. This would describe the twin scenario, one staying on their common life spiral, while one taking a shortcut between two points of the spiral. The twin taking the shortcut would live less of the joyful events occurring on the spiral timeline, but as we established earlier, would age less.

Similarly, we could envision traveling back in time on our time spiral, a topic we all but discarded as impossible. Our visualization system would certainly allow one falling back to an earlier spiral cycle of the person's life spiral. That would be in the same fashion as jumping forward, assuming some kind of a magic counterpart of time dilation like time contraction.

For example, one can imagine that the bio-medical industry develops methods of saving an imprint of our mind at certain stages of our lives. Then cloning ourselves (that is almost in the realm of science) at any other time in the future and restoring our mental status would enable us to retain an earlier physiological version of ours. We will leave the issue of the distinction between mind and soul to the reader.

But, did we really reverse the time in either case? The disappointing answer is not. In the case of the twin dilemma, the traveling twin will arrive home truly younger then the stationary twin, but sometime in the future. The bio-medical manipulations would extend the cycle of our life to a longer duration into the future. The emphasis is on the future. Neither scenario would really reverse the arrow of time that always points to the future.

The irreversibility of the arrow of time has some physical analogies. An example is complex physical systems, like alloys. They are created by melting and mixing components into a new material. The new material is more complex (at least in the sense of containing more components) and very likely much more useful in its physical characteristics. Once the alloy has been created, the components cannot be recovered.

Another example of time's irreversibility is based on the observation that nature seems to be always moving from a state of order toward disorder. For example, seeing a glass falling onto the floor and shattering into many pieces is a common occurrence, however, one never sees the pieces jumping up from the floor and assemble into the intact glass.

The constant and irreversible increase of complexity or disorder of physical systems mimics the behavior of time. The increasing complexity is manifested in the constantly increasing amount of history. The increased disorder is manifested in events of history that cannot be undone, like wars and natural disasters. The orientation of these is one directional, just like the arrow of time always points to the future.

There seem to be a possibility for the time arrow pointing back in the scale of cosmological events. According to this philosophy, the time arrow is pointing forward only when the universe is in its inflationary state that we are supposedly in at the moment. Once the expanding universe reaches the limit of its expansion and starts its deflationary state, the time arrow will be turning back.

We are forced to conclude with the understanding that we cannot conquer time and it is a futile effort to try to extend our life to go on forever. Unfortunately, at a certain stage we cease to exist as human beings and our Earthly bodies as a vessel carrying our soul will decompose. We will, however, exist in the future in our descendants' memories as they will be in theirs. A small consolation, but after all, we may live forever.

Epilogue

We arrived at the end of our cycle of time in this discussion about the cycles of time and completed our inquiry into this beguiling topic. We noticed that time has a very strong trichotomy of past, present and future. Their definitions are rather simple, the past we remember, the present happens now and the future we anticipate.

The turning point of time is the present separating two distinctly different regions of time. The past is definitive, its events have transpired and there is nothing we can do about it. Or can we? We explored the faint possibility of time travel, but it still seems like science fiction.

The future has a contrasting indefiniteness to it, we do not exactly know what will happen until it becomes present. Then it happens and it becomes past. Hence the present is a transition point from indefiniteness to definiteness. It appears that history marches through time, inevitably, unstoppably, but very measurably.

This moment of reading this sentence is our present and it puts the prior chapters into the past. The immediate future of reading the rest of this epilogue will cause this moment also to become the past. On this human scale time is a finite collection of life's moments.

On the universal scale time is an infinite collection of cosmic moments and celestial cycles. Since we found that the temporal distance between two physical events is relative, measuring their distance with time is not the only possibility. We could also measure their distance by their changing position with respect to our cosmic location in the 26,000 year long precession cycle.

One could say: "I was born when North was 5 degrees 30 minutes and 45 seconds to the right of Polaris". That would be somewhere in the second half of the last century. Note that the minutes and seconds here are not time units, they are lower units of the degree of an angle, they are spatial units. This is the final proof that time and space are inseparable.

Literature

[1] Barbour, J.; The end of time, Oxford University Press, 1999

[2] Barnett, J. E.; Time's pendulum: From sundials to atomic clocks, Harcourt Books, 1998

[3] Callender, C.; Introducing time, Totem Books, 2005

[4] Dale, R.; Timekeeping, Oxford University Press, 1992

[5] Duncan, D. E.; Calendar: Humanity's epic struggle, Avon Books, 1998

[6] Galison, P.; Einstein's clocks, Poincare's maps, Norton, 2003

[7] Hawking, S.; A brief history of time, Bantam Books, New York, 1988

[8] Michels, A. K.; The calendar of the Roman republic, Princeton University Press, 1967

[9] Richards, E. G.; Mapping time, Oxford University Press, 1998

[10] Steel, D.; Marking time, Wiley & Sons, 2000

[11] Tedlock, B.; Time and the highland Maya, University of New Mexico Press, 1992

[12] Waugh, A.; Time, its origin, its enigma, its history, Carroll & Graph, 2000

Index